Microcartography

MICROCARTOGRAPHY:

applications for archives and libraries

Edited by

Larry Cruse

with the assistance of Sylvia B. Warren

WESTERN ASSOCIATION OF MAP LIBRARIES

Occasional Paper No. 6

1981

Library of Congress Cataloging in Publication Data
Main entry under title:

Microcartography, applications for archives and libraries.

(Occasional paper / Western Association of Map Libraries ; no. 6)
"Outgrowth of a meeting held by the Western Association of Map Libraries ... at the University of California, San Diego ... October 6-7, 1977"--Introd.
"Annotated bibliography, by Larry Cruse": p.
Includes index.
1. Map collections--Congresses. 2. Microphotography--Congresses. I. Cruse, Larry, 1944- . II. Warren, Sylvia B., 1949- . III. Western Association of Map Libraries. IV. Series: Occasional paper (Western Association of Map Libraries) ; no. 6.
GA192.M52 025.1'29 81-19718
ISBN 0-939112-07-8 AACR2

Printed in the United States of America
at Santa Cruz, California

Information regarding this series and other publications of WAML may be found on p. 199.

Foreword

"Microcartography" is a new term, coined by Larry Cruse, to mean "all cartographic materials which have been photographically registered on film in such a way as to require magnification for their full utilization".

Cartographic information specialists have only recently been offered opportunities to enhance their collections with heretofore unavailable cartographic materials - through the medium of microcartography. This collection of papers documents the complexities of this yet evolving medium.

In his Introduction, Larry Cruse observes that "...there is a wide range of choice and plenty of avenues for ingenuity to explore". Indeed, it is because of the choices, and the rapid technological changes taking place in this field, that the Western Association of Map Libraries takes this opportunity to present this Occasional Paper. It is our hope that it will offer a few landmarks from which the librarian can steer a safe and successful journey into the field of Microcartography.

We are grateful to Larry Cruse for the expertise and concern he has developed in this subject. In addition to his overview, presented in his Introduction, his Annotated Bibliography will guide the reader beyond the experience of the presentations herein. Entry number forty-five refers to his presentation that has since been published in Bulletin No. 120 (June 1980) of the Geography and Map Division, Special Libraries Association. It is a landmark contribution to the literature of cartography, and leaves no doubt as to his expertise and leadership in this field. His continuing contributions to this speciality appear as a regular feature in the Information Bulletin, Western Association of Map Libraries, beginning with Volume 11, Number 2 (March 1980).

We also extend our thanks to the other contributors of this Occasional Paper, they are the pioneers without whom Microcartography would remain misunderstood.

In addition to the thanks extended to Sylvia Warren by Larry Cruse in his Introduction, we are indebted to Philip Hoehn, Chair of the WAML Publications Committee, for editorial assistance.

Stanley D. Stevens
Editor of Publications
Western Association of Map Libraries

TABLE OF CONTENTS

Foreword, by Stanley D. Stevens v

List of Illustrations, Figures, and Exhibits ix - x

Introduction, by Larry Cruse 1

Maps and Charts on Microfilm??, by Charles J. Barkee 9

Precision Navigation Filmstrips, by Thomas S. Fraim 17

The Space Imagery Center: a Multiformat Collection with a Computer-Linked Microfiche Retrieval System, by Gail Georgenson 27

Microcartography at the U.S. Geological Survey, by Gerald L. Greenberg 35

A Simple Map Microfilm Program, by R. Philip Hoehn 47

The Historical Maps of the City Engineer, City of Los Angeles: a Microfilm Collection, by Bernice Kimball 53

A Discussion of Microfilm Formats and Some Newly Developed Readers, by William R. Lataday 69

The Microfilm Programme of the National Map Collection, by Gilles Langelier 85

Application of 35mm Aperture Card Technology to Microcartography: How to Acquire a System, by Carl Lundquist 97

A Systematic, User Oriented Plan for Topographic Maps on Microfiche, by Robert Markham 109

The Standards Committee on Cartographic Microforms - a Progress Report, by Stanley D. Stevens 123

Commercial Reprographics Services, by Bill Talbot 129

The Manuscript Map Microfilming Project of The National Archives, by Charles Taylor 135

Maps and Microimaging, by John A. Wolter 143

Annotated Bibliography, by Larry Cruse 165

Index to the Bibliography, by Larry Cruse 195

LIST OF ILLUSTRATIONS

FIGURES, AND EXHIBITS

Microfiche Format (Figure 1a - Barkee) 14

Sample to accompany Barkee's "Maps and Charts on Microfilm??" (Figure 1 - Barkee) .....Inside Rear Cover

Reader for Boats (Figure 2 - Barkee) 15

Realist's tri-lens Vista-Viewer (Figure 3 - Barkee) 16

Geographic Area Relationships for Typical "A" & "B" Formats (Figure 1 - Fraim) 24

ONC A-1 Matrix (Figure 2 - Fraim) 25

Projection "Chunk" (Figure 3 - Fraim) 26

Space Imagery Center (Figure 1 - Georgenson) 33

Planetary Data, Lunar Data, Future Acquisitions (Figure 2 - Georgenson) 34

NCIC's Catalog System (Tables - Greenberg) ....44 & 45

Index Card & Aperture Card (illus. - Kimball) 62

Sample Check Sheet for Manuscript Maps (Kimball) .. 63

Sample Sheet from Historical Index (Kimball) 64

70MM Latady Camera (Figure 1 - Latady) 75

70MM Latady Camera technical data (Figure 1a - Latady) 76

Universal Precision Copy Camera (Figure 2 - Latady) 77

" ... technical data (Figure 2a - Latady) 78

Latady Precision Copy Lens (Figure 3 - Latady) 79

Latady Microview (Figure 4 - Latady) 80

Latady Microview (Figure 4a - Latady) 81

Latady MiNimap Viewer (Figure 5 - Latady) 82

Latady Viewer for 70 x 100mm film chips (Figure 6 - Latady) 83

Latady 105mm microfiche viewer ....(Figure 7 - Latady) 84

LIST OF ILLUSTRATIONS, FIGURES, AND EXHIBITS (continued)

Architectural Drawing (Illus. 1 - Langelier) 87

Mappemonde ... Couel ... 1765 (Illus. 2 - Langelier) .. 87

Sayward District, B.C., 1913 (Illus. 3 - Langelier) 89

Le Canada ... Duval ... 1664 (Illus. 4 - Langelier) 89

Seattle Special 1:62,500, 1909 Edition (Wolter) 144

Color slide, catalog card and shelf list card (Exhibit 1 - Wolter) 147

John Speed's 1676 Countie Westmorland (Exhibit 2 - Wolter) 148

Map Mounting and 35mm Color Filming Costs (Table 1 - Wolter) 149

Microfilm Equipment and Accessories Costs (Table 2 - Wolter) 153

Flow Chart (Exhibit 3 - Wolter) 157

35mm Microfilming Costs (Table 3 - Wolter) 158

1

Introduction

BACKGROUND

Microcartography is the outgrowth of a meeting held by the Western Association of Map Libraries (WAML) at the University of California, San Diego (UCSD), October 6-7, 1977. WAML's Publications Committee felt it was of sufficient general interest to map curators and librarians that a permanent record should be attempted. Since speakers had not been asked to submit formal papers for publication before the meeting, they were retroactively solicited. For those unable to prepare their own, tapes of their talks were transcribed and edited to provide substitutes. To make this distinction clear, such papers have been labeled "interpreted by Larry Cruse." Additional papers were either solicited or discovered and were included to add substance to the title's promise. For that reason, this is not strictly a proceedings volume of that meeting.

DEFINITION

The papers included were invited when "microcartography" meant only the photographic reduction of maps onto film. Those images were then intended for projection and viewing. But, limited to that context, a new term is not needed; either "map microforms" or "cartographic microforms" would have served just as well.

It was only after the meeting, when time could be devoted to the literature on this topic, that it became apparent the alternative labels were inadequate. "Map microforms" does not implicitly include remote sensing, and, even though "cartographic microforms" does, it does not adequately cover the diverse range of products and applications for which the film medium is suited but which are not strictly "cartographic" in nature, such as gazetteers, indexes and or text, or microfilm as a storage medium for computer data.

In addition to the inadequacies of the terms "map" and "cartography," there are also the conventions of micrographics, related by Stan Stevens in his paper on standards, which are not going to cover all of the possibilities, although they provide an essential foundation. For instance, there is the matter of microform and image sizes, many of which will be at variance with standard microform products: microcartographic products currently range from about 3mm up to about 241mm, extending beyond conventional microforms sizes at both extremes.

"Microcartography", then, refers to all cartographic related materials registered on film in such a way as to require magnification or enhancement for their full utilization. This includes maps, data sets, remote sensing imagery and supporting text. It includes all film formats, all film types which offer high resolution and the study of such materials in terms of human factors.

This volume limits itself to photographic reduction of paper maps and the generation of indexes to them; microcartography also includes maps never available in paper copy. This volume limits itself to Computer Output Microfilm (COM); microcartography also includes Computer Input Microfilm (CIM). Finally, microcartography includes the design-influence microforms are having on the presentation of maps and that maps are being designed now, at least in some cases, with this influence taken into account.

SIZE

The WAML meeting had three principal points of emphasis: format size, the use of color, and microform equipment (hardware).

Microform size is a strongly held issue, with irreducible merits assignable to each of the standard sizes, 16mm (Georgenson, Greenberg), 35mm (Fraim, Greenberg, Lundquist), 70mm (Barkee, Latady), 105mm (Langelier, Taylor), 150mm (Fraim), and hybrids of one format carried on another (Barkee, Georgenson,

Greenberg, Hoehn, Lundquist, Markham).

A curator who must deal with all of these potential microforms can either approach them severally or attempt a least common denominator approach. In the former case, housing the various formats will be troublesome since it is impossible to predict how many of which style cabinets or drawer inserts will be necessary. Trying to take a least common denominator approach involves a great deal more labor in reformatting the material, as Phil Hoehn illustrates, but yields savings in every other conceivable way. Its one limitation is confronted when a microform larger than 105mm must be accessed. At this point, the 105mm least common denominator is very similar to Carl Lundquist's 35mm LCD*, resulting in a circular problem. The only consolation is that a 105mm LCD reduces the problem greatly since it will carry all sizes up to 4" x 6". The only materials currently available in a larger format which can still be considered micro in nature have to do with remote sensing. So, if remote sensing materials are not a consideration, a standard 105mm carrier, interspersed with microfiche, should provide for all contingencies. If remote sensing has to be kept in the same facilities as the other microforms, it will have to be filed separately, segmented (cut into pieces), or photoreduced, in somewhat the manner as Gail Georgenson discusses in describing a graphic index.

COLOR

The second issue permeating the original meeting and these papers had to do with the unavailability of color microforms. The general conclusion was and is that color film is essential, at least some of the time, that it is inferior to black-and-white film in resolution and longevity, and that color processing is a problem. Although the expense of color film is mentioned, it is only compara-

* LCD = least common denominator

tively expensive vis-a-vis black-and-white microfilm. It is not prohibitively expensive when its space and labor savings trade-offs are taken into account. At the time of the WAML meeting, 105mm color transparencies could be delivered in volumes of 100 examples at a price of $3.00 each; 35mm would be 50¢ or less.

All who discussed this issue felt sure that some new development in color film technology would make its appearance in the near future. Since those sentiments were expressed, there has been a gradual evolution in the use of standard color microforms, which should at least make processing more practical, and a revolution, using embossing and light diffraction rather than conventional techniques (see Gale, 1978a,b, in the bibliography section). But such a revolution can only be brought about by a volume producer, such as a national mapping agency, not by individuals.

EQUIPMENT

As Carl Lundquist points out, when considering microforms, equipment availability is as important as any other consideration. Next to the need for cheap, durable, high resolution color microforms, the next most persistent demand is for an inexpensive, universal reader/printer. This could take one of several approaches, separately or in combination: zoom lenses, fiber optics, solid state arrays and/or high resolution TV monitors. In the mean time, adequate 35mm equipment begins at about $100 and multi-function 105mm viewers are in the $3,000-and-up class. Printing is another case entirely; most printing is 35mm based, is designed for engineering drawings and is fairly expensive. Equipment which can handle 105mm and all lesser sizes, starts in the $11,000 range. In-between (and beyond) are any number of hardware options, the choice of which is highly dependent upon individual circumstances. For instance, in a large library cartographic microforms can be integrated with other programs, such as slides and general microforms, to justify the purchase of multi-use equipment, including special electrostatic copiers, or, at least in the meantime, the copying burden can be left to patrons, who have access to commercial copying such as that mentioned by Bill Talbot. So, actually, there is a wide range of choice and plenty of avenues for ingenuity to explore.

ECONOMICS

Implicit or explicit in all of the papers is the role money plays in the decision to use microforms at all. This is so fundamental a consideration that it is often left unexplored. In this volume, Carl Lundquist provides data which will be useful for those calculating storage costs, while Robert Markham, Charles Barkee and Gerald Greenberg all point out the initial economies of purchase.

Translated into map library terms, a medium sized collection of 100,000 to 150,000 maps reduced to a micro format of 105mm or less would yield a savings on the order of $500,000 in terms of floor space and equipment not needed for an equivalent paper map collection. Further, such items as remote sensing are proportionately cheaper in microform (35mm vs 241mm), on the order of 15¢ per image as opposed to $3.00 - $25.00. With this kind of head start, a great deal of contract work can be underwritten, as well as some very exotic equipment with which to view and reproduce the result. And the savings are cumulative since labor is less with microform equivalents, up to and including complete automation of the collection.

HUMAN FACTORS

It is beyond debate, as Robert Markham points out, that paper copy is infinitely to be preferred to microforms. Only after a full day of reading a projected image can the initial bias be fully appreciated. Fortunately, few of us are consigned to such a fate.

Even so, the proportion of time spent viewing such images gives proportional appreciation: it will remain the curator's responsibility to do as much as possible to make the medium a positive experience. This includes arguing for the best optical equipment available and for the creation of a suitable environment. This situation is analogous to the environment in which navigational film strips are used, referred to in human factors studies as a "critical use situation." While the map curator will not have to design a working environment on this order, others have--only after studying the map/mortal relationship at length. Fortunately for mortals, their studies conclude that map redesign for critical use situations is the most practical approach to an improvement in information transfer. Such improvements, in turn, will make information transfer in non-critical use situations much more efficient. In terms of microcartography, this is the information-feedback-loop which will lead to improved map design generally, but especially in regard to microforms.

BIBLIOGRAPHY

An annotated bibliography has been appended to the papers to provide the interested reader with a sampling of work already done by others. To a large degree, it is built on the excellent foundation laid by Ralph Ehrenburg while at the U.S. National Archives, and the research and development work of the U.S. Army Engineer Topographic Laboratories at Ft. Belvoir, Virginia.

FUTURE TRENDS

It seems only natural for those who read the following papers to ask themselves how long microcartographic materials will be of use before they are superseded by computer cartography, or some other high technology. The fact of the matter is that no one really knows. Both micrographics and digital cartography are maturing simultaneously, but it is significant to note that they are compatible technologies, not mutually exclusive ones. At present, it is their mutual capabilities which are receiving emphasis. If this trend continues, any prediction for the future that map/remote sensing collections will be exclusively one or the other technology would be difficult to support. It appears more likely that curators and producers will add them to their existing repertoires so that we can look forward to having various "maps" in a number of formats, each used in a particular or even unique way. Thus, while the information may be redundant, its utilization will not be.

APOLOGIA

My apologies to those who have patiently waited for this book to be assembled. At the outset I knew I could not write, suspected I could not transcribe, and only lately realized that editing was not my forte either. Thus, my thanks go to Sylvia Warren, who has successfully neutralized most of those weaknesses in print. Without her help, it would have taken at least another generation.

La Jolla, California

November 1979

9

Maps and Charts on Microfilm??

Charles J. Barkee, President, Microchart Systems Incorporated, 819 A Street, San Rafael, CA 94901

"Maps and charts on microfilm? You gotta be crazy!" While seldom stated so blatently, it is not a totally uncommon thought on the part of many when first introduced to the subject. And I must admit, there certainly have been times when I felt the whole venture might be a persuit in madness. What my banker, attorney, accountant and wife thought is not relevant...and probably not repeatable.

Nonetheless, the product does exist; and with the passage of time it appears the concept is not only being accepted but, in numerous environments, readily embraced. The reasons behind its acceptance are, to put it very simply: It fills a need...in fact, a number of them!

Libraries, commercial firms, educational institutions and just plain people do have problems with paper charts and maps. They have difficulties in storing them (SPACE) and problems retrieving and refiling them (STAFF TIME). They're concerned about the durability of paper charts (PRESERVATION); and certainly an increasingly major concern, they're having problems justifying the expense of procuring maps and charts (COST)--particularly for those areas which don't have a high reference factor. They have found that all of these problems can be significantly reduced through the use of microfilmed cartographic materials.

Let me state at the outset, we harbor no beliefs that microfilm will ever replace paper. There are many things you just cannot do on film. BUT, unequivocally, microfilm can solve or help to reduce the SPACE, REFERENCE and COST problems, plus (and this is proving to be a prime benefit) allow the user to

EXPAND his map and chart coverage.

Today our firm, Microchart Systems Incorporated, is serving customers in many parts of the world..from commercial fisherman off the coast of Newfoundland ...to university libraries in California...to commercial marine and oil exploration firms with interests in the North Sea and the Persian Gulf. It is a viable and valuable product for all of these applications.

Quite possibly, the how's and why's of our company's formation, plus the problems faced, would be of interest. In 1969 a retired Honeywell engineer, Mr. DeForrest Olson, started the firm as Microchart Navigation Systems (MNS) in Seattle, Washington. The original marketing thrust of the business was pointed directly, and exclusively, toward the commercial fishing fleet based in the Pacific Northwest, British Columbia and Alaska. The only product considered at the time was the nautical chart--and only as an aid to navigation. Hence, the word "Navigation" in the original company name.

The market direction seemed sound enough as charts were a problem for the fisherman. The typical commercial fisherman working the waters from Seattle to the Cook Inlet of Alaska was faced with purchasing, storing, retrieving and refiling several hundred nautical charts. Additionally, the space allotted him for the storing and the using of the charts was a major concern. And, as if to compound the situation even further, the government's chart-creating agencies (as is their responsibility) revised the charts with great regularity. This left the fisherman with an uneasy feeling that, should he not have the most current chart and have a boating mishap, he was leaving himself wide open to legal recourse. However, even with this market that so desperately needed a new approach, the problems Mr. Olson faced were formidable. There were the technical problems, such as which microfilm format to use, roll film, aperture cards or microfiche? What type of camera would handle the multi-color and indefinite details of nautical charts? What reduction ratio, type of film, and which readers or display devices would fit into the stringent space requirements aboard a pitching, rolling vessel?

Equally pressing was the HUMAN problem. Would the fisherman, by nature a conservative and tradition-bound individual (with a capital "I") accept such a radical change as nautical charts on microfilm?

By 1973 most of the problems had been worked out. The format selected was microfiche because of its superior handling and flexibility features (Figure 1; sample microfiche inside back cover). Decisions regarding the very special cameras had been reached after what seemed like endless wrong avenues. Types of film had been tested, and silver positive was finally selected. The proper reduction ratio was determined. Readers were selected and modifications for off-shore/on-boat use had been engineered (Figure 2). Even the "human" problems seemed to be working out. The company began to actively market its product. By 1975 MNS had several hundred fisherman on both U.S. coasts utilizing the system. There were minimal dollars available for advertising, but the customer list was experiencing fine growth, thanks to an effective user word-of-mouth campaign. Unfortunately, the long hours and efforts of developing a company from-the-ground-floor-up was taking its toll on Mr. Olson. In mid-1976 he suffered a fatal heart attack. While MNS certainly did not fold, the loss of Mr. Olson's drive and enthusiasm did affect the company's growth. Expansion plans were curtailed, new product plans cancelled and, in general, a slowdown ensued. Finally, the business was put on the market.

In late 1977, my wife and I purchased the company from Mr. Olson's widow. My personal background consists of nearly twenty years of microfilm sales and marketing management positions with several major microfilm manufacturers. We continued to operate the business from Seattle until early 1978. At that time we decided to bring it closer to our home in California and moved the headquarters to its present location in San Rafael, a suburb north of San Francisco. The commercial fishing market continued to be our major target, with a smattering of promotion directed toward the ocean-going pleasure boater. Growth, while adequate, was not up to our expectations...or our effort expended. As the saying goes, "we racked our brains" attempting to find the key to growth and profitability.

As luck would have it, we found part of the answer one rainy night in Denver. While in that beautiful Colorado city on a consulting job, I planned to play nine holes of golf after finishing my assignment late in the afternoon; this plan was rained out by a violent thunderstorm. The evening's remaining options ran the gaumet of watching TV reruns at the hotel to an evening at the Denver Public Library. The more I thought about watching Gilligan's Island and Maude, the better the library sounded. And so while browsing through the Map Section, I noticed a chap studying a paper nautical chart. This rather dumbfounded me, as I hadn't envisioned anyone in Colorado needing a nautical chart, let alone a library having one.

To the contrary, the Map Section clerk informed me that they indeed did have nautical charts and, in fact, they were used quite frequently by the patrons. She went on to state they would like to carry more but space and cost were definite problems. Well, bells rang, whistles blew and lights went on! Here was the start of a new market. And to make the story short, succinct, and successful, the concept has evolved into the development of what we refer to as the "On-Shore" market.

Today this market includes numerous college/university libraries, commercial marine transportation companies, major oil and gas companies, and government agencies. A major factor in the success of this market breakthrough was our decision to package hardware with the software as an option to facilitate the "getting into" of the system (Figure 3). In addition, we developed a unique package library concept that greatly simplified the ordering of a large number of charts for a given area. Still another change was that serving this new and larger customer base required that we greatly expand our chart coverage. Today, we provide key coastal charts for ALL regions of the world. It's probably a poor slogan for someone in the nautical chart business, but it was "sink or swim." We chose to swim.

Still another stroke of good fortune occured in 1978...and again, it happened in Colorado. It was at this time we made contact with Dr. Bob Markham. Bob is head of the Micrographics Laboratory at the University of Northern Colorado (UNC) in Greeley, Colorado. The Micrographics Laboratory's stated charter is to develop useful and unique applications utilizing microforms. One of those applications was the microfilming of topographical maps. The problems confronting the large volume user of "topos" were strikingly similar to those of nautical chart users. And the Micrographics Laboratory of UNC, like Microchart Navigation Systems, experienced early problems with format and quality. Through extensive effort and much patience, Bob and his staff solved the technical questions. However, one major problem remained. Being an educational institution, the Micrographics Laboratory was seriously limited in its ability to market its product. Additionally, the joint marketing of hardware with software was completely out-of-the-question. Conversely, Microchart Systems Incorporated, as a commercial firm, had no such restrictions. The outgrowth of all this is an exclusive agreement for our firm to market the University of Northern Colorado Micrographics Laboratory's topographical microfiche products.

Needless to say, we are extremely optimistic about the potential that microfilmed topographic maps offer---to the user, UNC and MSI. The market appears virtually unlimited.

So that is where we are today, producing government-created nautical charts and topographical maps on high quality silver microfiche. Space savings, faster reference, and a more durable map or chart are supplied at no extra charge, and all represent solutions to the very real problems mentioned earlier. Also, new laws such as California's Proposition 13 are placing increasing emphasis on the reduction of expenditures; microfilm can definitely be of service there.

As for our future plans, while we cannot divulge all of them, it is no secret that we are continually expanding our map and chart coverage. We are examining the feasibility and potential of utilizing our special camera equipment, technology and filming expertise for rare map/chart collections. A number of influential map curators have approached us on that. Also, we are designing several new and very attractive hardware/software packages built around some special market needs. And, coming full-circle, we are developing a ship-board reader for the offshore market that we feel will become a standard piece of pilothouse gear within a decade. With such large ambitions, you can see why we say, "at Microchart Systems only our product is small!"

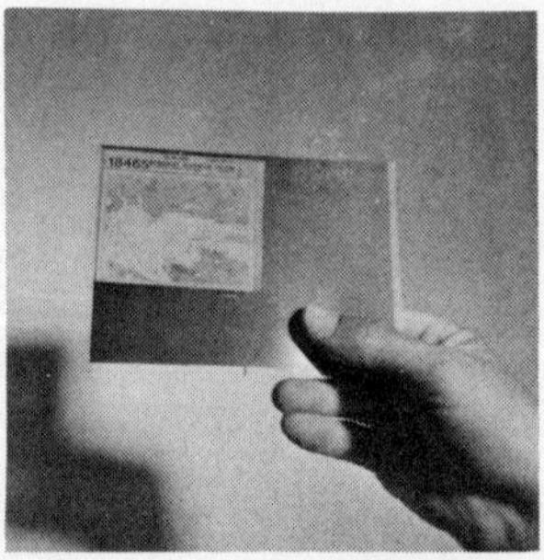

Figure 1a: Microfiche Format

(See sample attached to inside of rear cover.)

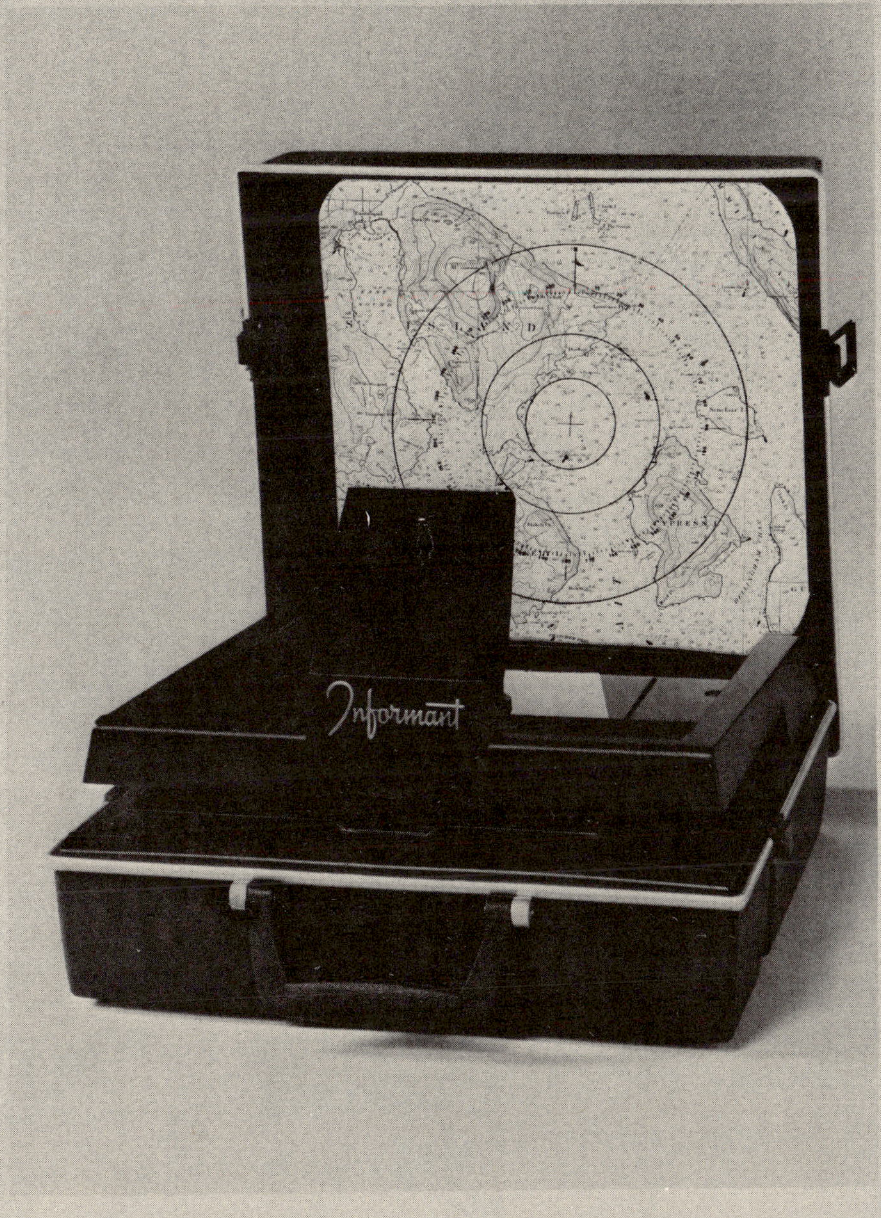

Figure 2: Reader for Boats

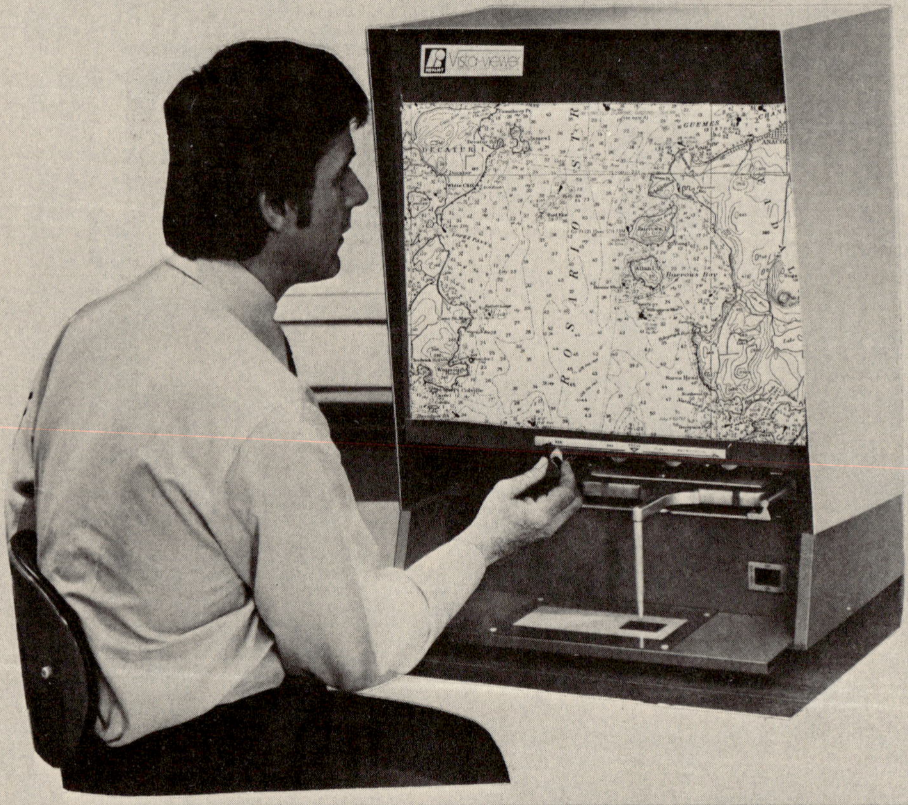

Figure 3: Realist's tri-lens Vista-viewer

Precision Navigation Filmstrips*

Thomas S. Fraim, Directorate of Plans and Requirements, Defense Mapping Agency Aerospace Center, St. Louis Air Force Station, Missouri 63118

ABSTRACT

The Defense Mapping Agency has been producing high resolution precision navigation filmstrips since 1969. These filmstrips are used in various type projection systems within DoD aircraft to display real time aircraft position. The Defense Mapping Agency Aerospace Center (DMAAC) developed the production procedures which required procuring new equipment, modifying onboard equipment, and developing new compilation techniques. DMAAC is currently producing filmstrips for test and evaluation in new aircraft systems which will require production of filmstrips into the 1980s.

BACKGROUND

In 1969 the Defense Mapping Agency Aerospace Center (DMAAC), formerly the Aeronautical Chart and Information Center (ACIC), was designated by the Defense Intelligence Agency (DIA) as the Department of Defense (DoD) filmstrip production agency. The initial requirement for navigational filmstrips came from the United States Navy in 1969. The filmstrips were to be utilized in the Projected Map Display System (PMDS) of the A7E aircraft. Shortly thereafter the United States Air Force submitted a requirement for navigational filmstrips for the Strategic Air Command (SAC) and the Tactical Air Command (TAC) F-111 aircraft.

* Presented to the First Pan American Symposium on Automation in Cartography, Pan American Institute of Geography and History, Washington, D.C., March 1978.

These two programs launched ACIC/DMAAC into the arena of precision navigational filmstrip production. Equipment was procurred and procedures developed and ACIC initiated delivery of filmstrips to the Navy in 1970. Since that time the Center has produced 32 operational A7 filmstrips, 13 operational F-111 filmstrips, as well as several test filmstrips for Air Force and Army helicopter programs. DMAAC is presently producing six (6) operational filmstrips for the Air Force HH53 helicopter Pave Low III system. Considerable cartographic and production expertise was required to solve the many complex problems involved in the new area of filmstrip production. For example, the small 35mm filmstrip size forced the Center to modify on-hand photographic equipment; higher film resolution requirements strained the state-of-the-art; and technical coordination was constantly required between ACIC, the user and outside contractors.

FILMSTRIP CHARACTERISTICS

The Projected Map Display System (PMDS) in the A7 aircraft, manufactured by Computing Devices Company, uses computer processed inertial navigation system data to furnish a continuous real time display of the aircraft's position. The DMAAC produced 35mm color filmstrip is the source of map imagery projected onto a 5" diameter ground glass screen. This system utilizes filmstrips produced from 1:500,000 Tactical Pilotage Charts (TPCs) and 1:2,000,000 Jet Navigation Charts (JNCs). A typical strip is 37 feet in length portraying approximately 800 x 1000 nm of TPC coverage and 1000 x 2000 nm of JNC coverage. The filmstrip contains approximately 200 frames with a filmstrip chart scale approximately 1/12 the source chart scale. The PMDS optics system projects the imagery onto the screen at a 16 time enlargement, thus producing a projected image scale approximately 1.4 times the source chart scale. The Army and Air Force helicopter systems incorporate PMDS' similar to the A7. The physical characteristics of the filmstrips are similar, however, due to different mission requirements the scale of source charts is different. The Air Force helicopter system uses 1:250,000 Joint Operations Graphics (JOGs) and 1:1,000,000 Operational Navigation Charts (ONCs). The Army system will utilize 1:50,000 topographic maps and 1:250,000 Joint Operations Graphics.

The F-111 aircraft has an advance avionics system which incorporates the Horizontal Situation Display (HSD) for projecting map imagery. As with the A7 system, the HSD system uses computer processed inertial navigation system data to drive the filmstrip in order to display aircraft position. An inherent component of the HSD is the DMAAC produced 150mm (6") color filmstrip containing chart images. The chart image is projected onto the screen of a cathode ray tube which has a special face plate for improving readability in a brightly lighted environment. The F-111 system filmstrips incorporate navigation charts at scales varying from 1:5,000,000 to 1:500,000. The 20 foot filmstrip contains 50 map frames at 1/7 the source chart scale for the 1:1,000,000 Operational Navigation Chart (ONC) and 1/10 the source chart scales for the other charts used. In the HSD the imagery is projected at a 10 time enlargement.

PRODUCTION PROCESSES

(The following discussions on production techniques will be oriented towards 35mm filmstrips. Keep in mind that procedures for 150mm filmstrips are similar.)

The production process is subdivided into three prime areas:

-Generation of master projections.

-Paneling lithographic copies of the charts to the projections.

-Photographing panels onto high resolution color film. The end result is to ensure compatability between the computer software which drives the aircraft map display and the chart/frame relationship on the filmstrip.

The first step in producing a filmstrip is to generate a projection compatable with the aircraft system software. For example, the A7 system uses a Lambert Conformal Projection, whereas, the Army helicopter system uses an Azimuthal Equidistant projection. The type projection specified is that which is best suited for the aircraft's mission. Also specified are various formats which define the geographical orientation of coverage for the various chart scales. These have either a predominant north-south orientation or a predominant east-west orientation. (Figure 1) Also specified for the A7/HH53 filmstrips is a reference point referred to as the LO/LAL which is defined as the intersection of the central meridian and the lowest usable latitude for the format selected. The LO/LAL ties the generated frame matrix to the projection. The frame matrix is defined by the number of frames east-west and the number of rows north-south. As an example, the Pave Low III ONC A-1 format has a frame matrix of one frame east-west and 25 rows north-south. (Figure 2) Other inputs are the frame dimensions at the filmstrip scale, spheroid values, filmstrip scale, compilation scale, and standard parallels for the projections. Based on these inputs, the DMAAC computer software generates a projection superimposed on the frame matrix for the geographical area defined. The data is stored onto magnetic tape which in turn drives an automatic plotter. The projection and matrix are plotted on stable base polyester film in a 13 x 41 inch frame format. Each projection, called a "chunk", is plotted three frames high. (Figure 3)

The next production process, paneling the lithographic charts to the projection "chunks", is a major portion of the filmstrip compilation. It is necessary to panel the lithos because of shrinkage in the non-stable paper used in printing, the frame matrix does not correspond to chart areas, and/or the lithos have been produced on different projections. The paneling, which involves cutting the lithos into various size pieces and then fitting these pieces to the projection, is accomplished to a .005 inch tolerance. The paneling may be thought of as either a "shrinking" or "stretching" of the lithos to fit the projection. The number of cuts required depends on the amount of "shrinking" or "stretching" required to maintain the .005 inch tolerance between the master projection and the chart projection lines. The cuts also distribute errors over a large area so that no gaps show between separate litho pieces. The litho pieces are paneled to the projection using wax as an adhesive, which allows a piece to be placed in its approximate position and eased into its exact location with little effort.

In those chart areas of white or very light blue, colored translucent paper is applied to reduce the light projected on the screen in the display system. Each chart used in the filmstrip is updated to include information potentially hazardous to low level flight. Of major concern are obstructions (towers, powerlines, and similar features) and significant cultural features, that did not exist or were not known to exist when the chart was produced. Various overlays are photographed onto each frame which indicate the outer edge of filmstrip coverage and the useful area of coverage for each frame. All compilation and photographic materials are prepunched prior to production for precise registration.

The final phase in the production process is the precise sequential photographing of the frame panels onto the film base. The film resolution, color densities, and frame position are very critical.

A Latady 70mm precision reduction camera with a nominal reduction factor of 12.053 is dedicated for all 35mm filmstrip work. The camera, with a Micro

Nikkor Lens, is a stock item. A specially built camera back, manufactured by Computing D vices Corporation, is used for precision film advance. The camera back can accommodate 100 feet of 35mm film and is capable of film advancing to an accuracy of .001 inch. The film is advanced manually based on settings computed by testing. The exact center of the filmstrip, designated the "A-Frame" is the reference point for computing advance settings. The setting for the "A-Frame" is determined and from this number a constant (22708.65), which corresponds to exactly 3.390 inches of film, is added or subtracted to obtain settings for each exposure. The exposures are butt joined with a .002 inch maximum overlap. Each exposure, corresponding to a double aperture exposure, covers one 12 x 41 inch panel containing two filmstrip map frames. Because the aircraft system "thinks" in terms of map inches, the placement of map frames on the filmstrip is very critical. Also, frame placement relative to the film sprocket holes must be maintained to very close tolerances. Very precise alignment between the panels and camera frame is required to accomplish accurate positioning. A three pin registration system, vacuum frame, a master alignment grid, and a series of reference marks are used. The frame alignments are checked on a Mann Comparator.

The film used is a very high resolution Eastman Kodak 35mm color film capable of 200 l/mm resolution. Color densities are controlled ($\pm$.05) by use of a 50% gray card and read outs on a densitometer using Status A-type filters. The light source is filtered at the copy plane in order to maintain maximum possible resolution and color balance.

Both the 35mm and 150mm film masters are produced in a Kodak 1811 Color Ecktachrome Processor. The 35mm duplicate filmstrips are produced using a 180 foot per minute continuous loop printer. A step and repeat type printer is used for the 150mm duplicates. The resolution specified for the filmstrip master is 120 l/mm and for the duplicates 85 l/mm.

For the F-111 filmstrips, a universal Latady camera with a GOERZ Lens (focal length - 12", f - 6.3) capable of 200 l/mm resolution is used. This system incorporates a Miller Holzworth 6" camera back capable of single frame manual advance utilizing a Geneva movement drive system. Nominal reduction factors of 7X and 10X are utilized in F-111 filmstrips. F-111 frames are butt joined but appear as separate frames on the filmstrip. The center sprocket hole of each frame is the reference point for positioning.

DMAAC is currently in the process of procuring a universal filmstrip generator capable of handling film widths from 35mm to 6.0 inches. This reduction copy camera system will utilize an HLC Dekacon III Camera System or equivalent and a universal roll film back. The camera back will be capable of handling film widths ranging from 35mm to 6.0 inches, perforated or non-perforated. The film drive will provide automatic metering for producing the filmstrips meeting the A-7, F-111, and future cockpit display system requirements.

CONCLUSION

The filmstrips produced by DMAAC have been enthusiastically received by the Department of Defense (DoD). They have proven so successful in various aircraft and all types of operations, that high resolution precision navigational filmstrip systems have become an integral part of aircraft display systems. DMAAC is currently working with systems analysts toward the design and future production of filmstrips for advanced aircraft under development.

ACKNOWLEDGEMENTS

The author would like to acknowledge the efforts of Mr. Charles Sullivan, Mr. Gene Stender, Mr. Jack Nesler, Mr. Richard Gephart, and Mr. Dick Sides, DMAAC, in the preparation of this article.

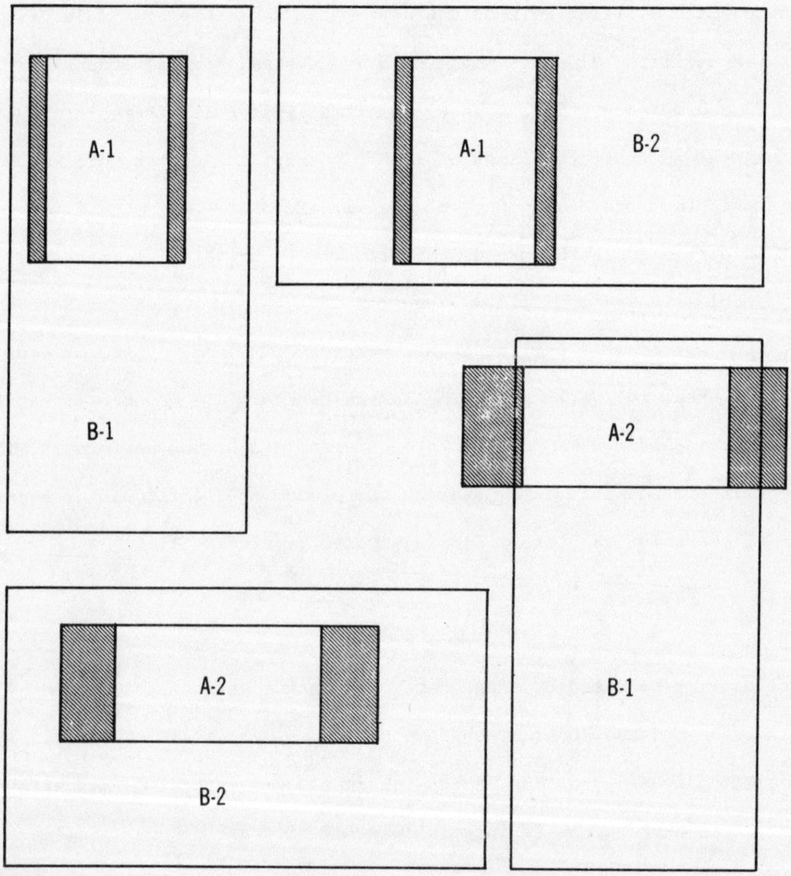

Figure 1: Geographic Area Relationships for Typical "A" & "B" Formats

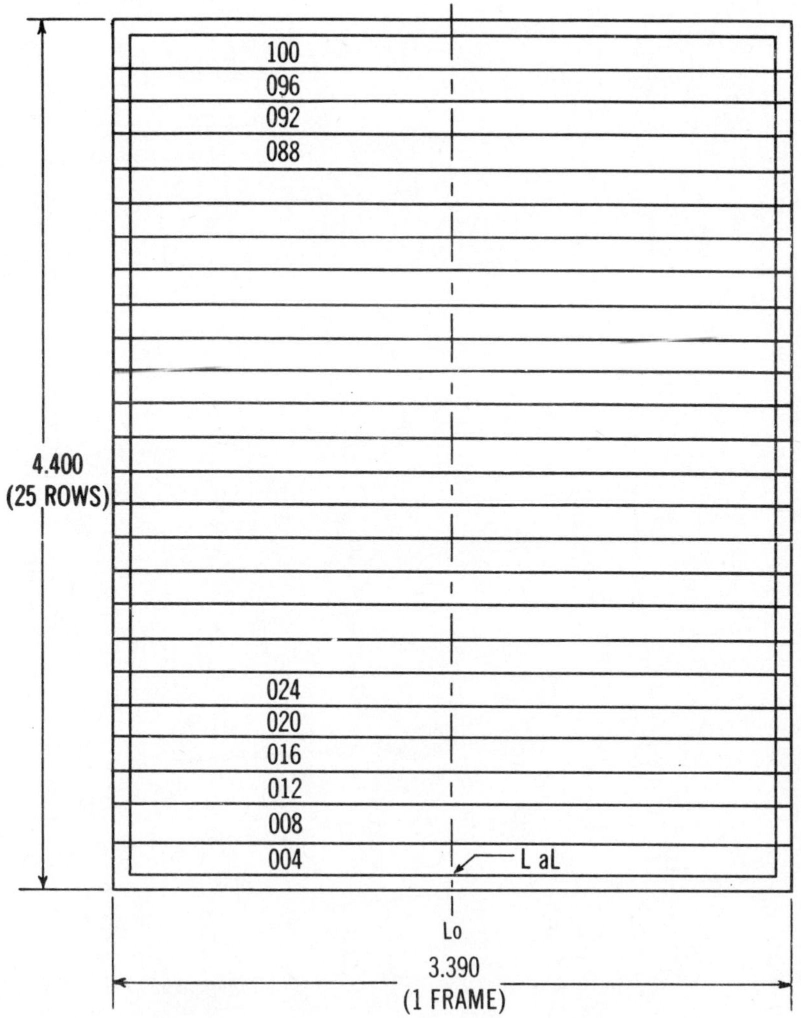

Figure 2: ONC A-1 Matrix

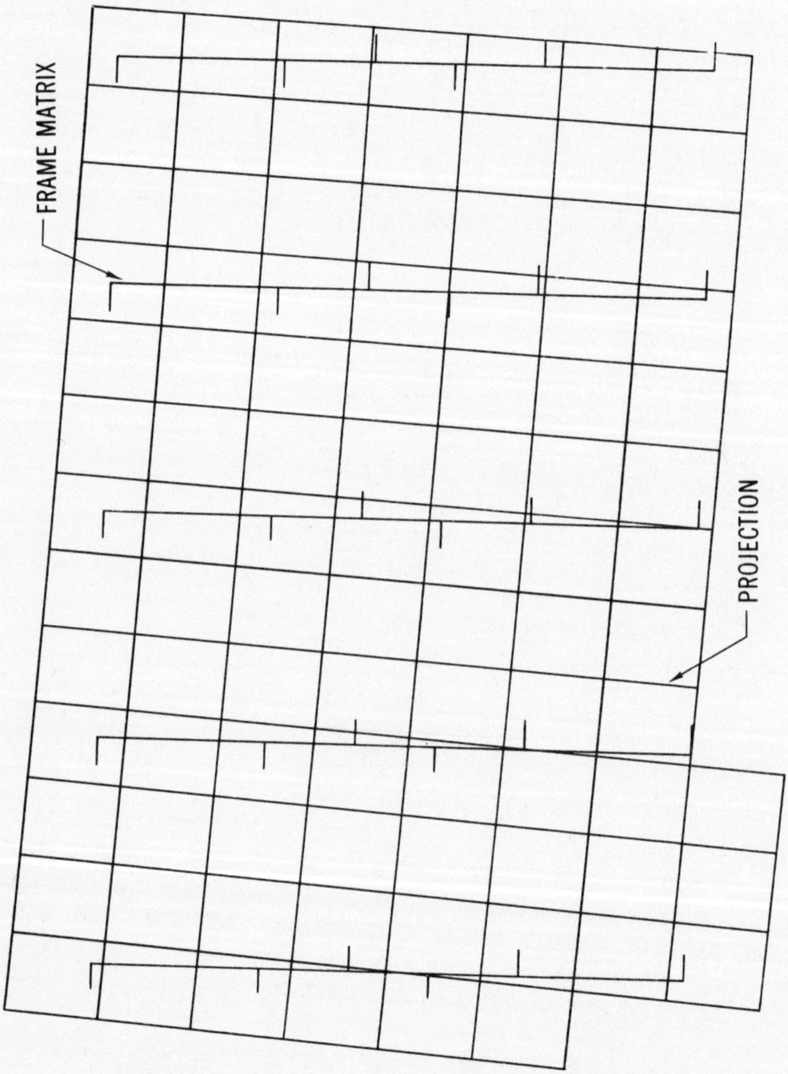

Figure 3: Projection "Chunk"

The Space Imagery Center: a Multiformat Collection with a Computer-Linked Microfiche Retrieval System *

Gail Georgenson, Librarian, Space Imagery Center, Lunar and Planetary Laboratory, University of Arizona, Tucson, AZ. 85721

As interpreted by Larry Cruse

The Space Imagery Center is a new collection having just opened June 30, 1977. It is an integral part of the University of Arizona's Lunar and Planetary Laboratory, where research involves such varied disciplines as geosciences, astronomy, chemistry, and physics. The Lab is closely associated with various NASA space programs, faculty members having served as principal and co-investigators on a variety of NASA's imaging experiments.

During the past 15 years, the Lunar and Planetary Laboratory has acquired a large collection of spacecraft and earth-based imagery of the moon and planets. This material comprises one of the largest and most complete collections of its type in the world. It includes imagery from every NASA imaging experiment flown to the moon and planets as well as the cartographic products produced from this imagery over the past 10 years. Also included in our collection are a large number of earth-based photographs of the moon and planets taken through telescopes. This is a unique collection and is unavailable elsewhere. On the other hand, the mission oriented imagery is available at several locations around the U.S. All members of a NASA mission imaging team receive the data produced by a mission.

*Presented to the Fall Meeting of the Western Association of Map Libraries, La Jolla, California, October 7, 1977.

With the successful completion of the Apollo program and the launching of the Mariner 9 and 10 missions to Mars, Venus, and Mercury, the collection grew so quickly that it was impossible to store and catalog the material adequately under prevailing conditions. Imagery was scattered throughout several offices, in closets, and under desks. In short, it was a real disaster. To solve the problem, a proposal was written to NASA to establish a center which would constitute a well organized, cataloged, working collection of lunar and planetary imagery, to be used as a research tool in the planetary sciences. A contract was awarded and I was hired in mid-1975 to pull the ever mounting data together and establish proper storage and retrieval procedures.

Today, the Center occupies approximately 1,100 square feet of floor space (An architectural rendering and floor plan are shown in Figure 1). We utilize a variety of different storage facilities to accomodate a range of formats that varies from 35 mm film strips to five-foot photomosaics. In between we have a number of 20 x 24 inch prints returned from the five Lunar Orbiter missions. They are printed on very thick paper and are quite heavy. Originally I planned to store them in hanging files suspended from rods. But they were too heavy and I ended up flat filing them in regular map drawers. Nor have I come up with any new ways to file film imagery, relying, for now, on traditional methods because the different types of format are too constricting. We also have 5 x 48 inch panoramic prints of the Lunar surface which are stored flat in a custom built cabinet. We use binders to store our 8 x 10 inch prints while 5 x 10 inch Viking pictures of Mars are stored standing up in file drawers (All data available for each planet is summarized in Figure 2).

Relevant supporting documents and notes issued by NASA, the Jet Propulsion Laboratory (JPL) at the California Institute of Technology (Caltech), and the National Space Science Data Center in Greenbelt, Maryland are at the Space

Imagery Center. We also have a very small selection of Landsat imagery of the Earth, for which we receive many requests and which we would like to augment.

Examples of the maps and atlases we hold include index maps to the mission oriented photography and film. We also have shaded-relief and geologic maps of Mercury, Mars, and the Moon produced by the U.S. Geological Survey. Tectonic and topographic maps of the moon are also in the collection. As researchers have more time to work with imagery, similar maps of this type will be produced.

Lunar and planetary atlases, such as the Consolodated Lunar Atlas and the Atlas of Mercury are also available. Work has begun on an atlas of Mars using the imagery returned by Mariner 9 and the Viking missions.

The data now contained in the Center represent only a beginning. Some 50 to 60 Viking Mars images are received in the Center each month. The two Voyager spacecraft now enroute to Jupiter and Saturn will return a possible 60,000 images of those two planets and their satellites in early 1978. Pioneer Venus will return additional photos of that planet. The question constantly confronting me is not simply where to store all of it (although that does cross my mind), but how to maintain some semblance of bibliographic control over it?

In regard to reducing all of this disparate material to a uniform microfilm format for study purposes, I am afraid that our scientists will always need to use the earliest generation of imagery possible. The loss of resolution resulting from an attempt to convert our material to microfiche would be unacceptable for study purposes. While this means that we will probably always have a multiplicity of formats, it does not preclude our microfilming a browse file to access the imagery.

The principal catalog access to the collection of mission oriented imagery is the computer-interfaced microfiche retrieval system. This system is only our access point, it does not in any way replace the hard copy imagery. It consists of a computer-linked Image Systems Incorporated microfiche viewer, a terminal, and an acoustic coupler. Microfiche used in this system is adapted by the addition of a metal clip along one edge which carries a unique binary identification code prepunched at the factory. This clip is used by the fiche reader to identify the desired microfiche. For instance, if a particular microfiche frame has the code 401-2F, the first element, "401", represents the microfiche on which the record is stored. The second part, "2F", represents the section on that microfiche where the desired frame is stored.

The fiche can be called up to the viewing screen in two ways. The first is a manual method involving push-button selection from the control panel and then automatic retrieval. To actually display the image, a "view" bar is then pressed. Since this manual method bypasses the computer completely it can only be used when the number of the desired frame is known.

The second method for displaying microfiche consists of routing requests through a computer. This method is more complicated but more useful, since users often do not know the image numbers they would like retrieved. It is used in all cases where prior experience and memory do not make the frame number familiar. Images are retrieved by imputing to the computer, via a terminal, any of a variety of paramaters, such as surface coordinates, topographic features, resolution, and the like. The computer searches its data files for images containing the user's specifications. Images which contain the intersection of these specifications are displayed one-by-one at the user's discretion. Data are printed out for the images displayed, if desired. Basically, that is the manner in which the system operates.

We have not been without our problems, both hardware and software. First, the mechanical problems. The carousel holds approximately 750 microfiche, which translates into 45,000 images in Cosati (Committee on Scientific and Technical Information of the Federal Council for Science and Technology) format. This presents a problem because, with 500,000 images to browse, we need to be able to store more than 750 microfiche. The problem would not be so acute if the entire carousel of microfiche was portable, but it is not. There are removable cartridges allowing for the removal of 52 fiche. Unfortunately, removing them leaves the adjacent cartridges unsupported, causing another 26 fiche to literally fall out of the carousel. For our purposes, it is a very poor design and needs modification to insure more versatile storage of the microfiche.

The microfiche itself was originally filmed on an acetate base medium. When Tucson's humidity drops, often to as low as 5%, the emulsion tends to dry out, causing the fiche to curl. Then, with the rise in humidity associated with our summer monsoon season, the fiche return to normal. So, while the low humidity is not permanently damaging, it creates a terrific nuisance during the winter months, jamming the retrieval unit and making viewing impossible. In fact, some of the encoded strips were actually torn from the microfiche. Probably the only real solution to this type of problem will be environmental control of the room in which the retrieval unit operates.

In theory, the computer program is a work of art, given the large quantity of data in the files. It is a program borrowed from Caltech, designed by a student who took the source codes with him when he left. Thus, we have been unable to add to the program and data files.

It would be a great help if we could convert the browse file to ultrafiche (microfiche reduced more than 90 times), rather than the 20x reduction microfiche we now use. But then all of our browse file would have to be readdressed and the resultant imagery would be a further qualitative compromise.

Mastering these problems while continuing to access proliferating imagery will be our challange for the immediate future. But we are still a new collection and hope to make improvements where we can. Hopefully, our experience with automatic retrieval equipment will prove of use to others designing their own systems, which will feed back to us, helping us to improve our own service. And, last but not least, we are always happy to share our expertise on space imagery, should you find yourselves faced with potential users.

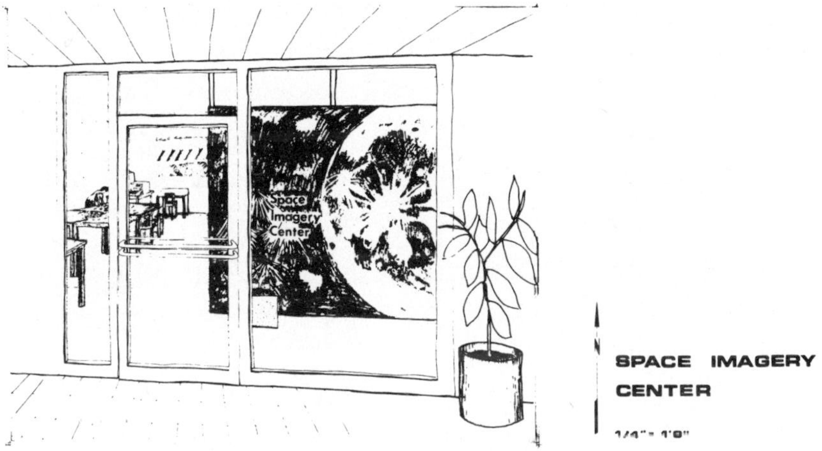

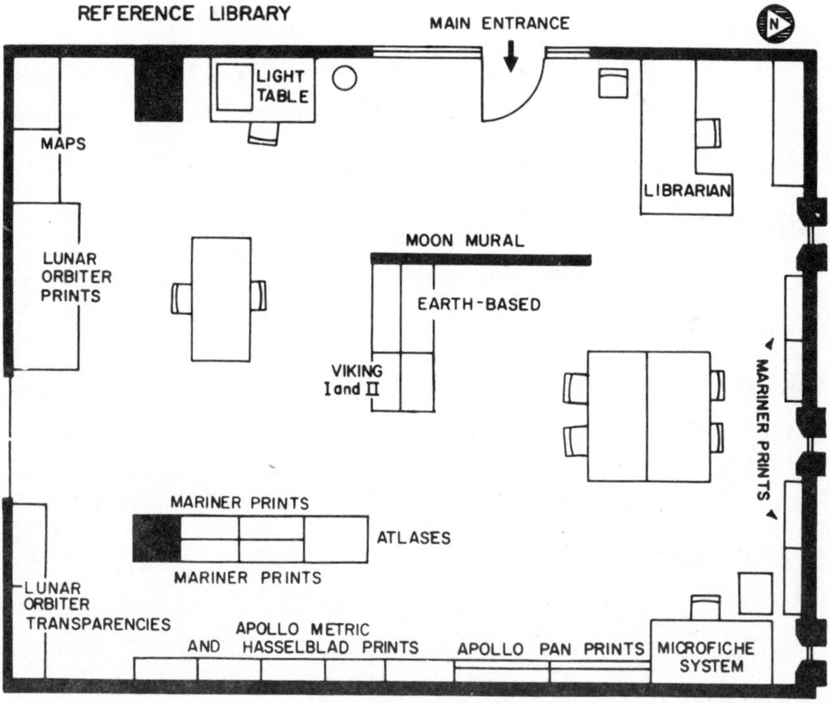

Space Imagery Center Floor Plan

Figure 1

PLANETARY DATA

Mercury	Mariner 10 (8 x 10 in. prints; 70-mm strip contact print stereo pairs; photomosaics
	Earth-based telescopic photos (35-mm)
Venus	Mariner 10 (8 x 10 in. prints; 70-mm strip contact print
	Earth-based telescopic photos (35-mm)
Mars	Mariner 4, 6 & 7 (8 x 10 in. prints)
	Mariner 9 (8 x 10 in. prints; 70-mm strip contact prints)
	Viking Orbiter (5 in. strip contact prints; duplicate negatives; photomosaics)
	Viking Lander (selection of prints)
	Earth-based telescopic photos (35-mm)
Jupiter	Pioneer 10 & 11 (8 x 10 in prints)
	Earth-based telescopic photos (35-mm)
Saturn	Earth-based telescopic photos (35-mm)
Uranus	Earth-based telescopic photos (35-mm)
Neptune	Earth-based telescopic photos (35-mm)
Comets	Earth-based telescopic photos (35-mm)

LUNAR DATA

Ranger 7-9	(8 x 10 in prints)
Surveyor 1, 3, 5, 6, 7	(8 x 10 in. prints; photomosaics)
Orbiter 1-5	(20 x 24 in. prints; 20 x 24 in. positive transparencies)
Apollo 8, 10, 11, 12, 13, 14, 15, 16, 17	(8 x 10 in. Hasselblad prints; 70-mm Hasselblad positiv transparencies; 5 x 5 in. metric camera prints; 5 x 42 in panoramic camera prints and positive transparencies)

FUTURE ACQUISITIONS

Voyager 1 & 2

Pioneer Venus

Pioneer 11 (Saturn)

Figure 2

Microcartography at the U.S. Geological Survey*

Dr. Gerald L. Greenberg, National Cartographic Information Center, U.S. Geological Survey, Western Mapping Center, 345 Middlefield Road, Menlo Park, California 94025

As interpreted by Larry Cruse

INTRODUCTION

I am going to discuss some of the uses of microcartography at the U.S. Geological Survey's National Cartographic Information Center (NCIC). Some of the systems described are in use, while others are going to be available to the public in the near future.

But first, a basic item: Why micrographics instead of other modes at the Geological Survey? Our reasons are common to most microform users: (1) low initial cost, (2) low cost of updates, particularly with regard to microfiche, (3) the wide availability of microform readers in schools, libraries, and public agency offices, (4) portability, (5) size, and (6) smaller storage space.

NCIC is currently using all available modes of microfilm, including 105mm microfiche--particularly Computer Output Microfiche (COM)--35mm silver or diazo roll film, 35mm aperture cards, and 16mm film cassettes.

RECORDS

Some of the general types of records NCIC is microfilming include actual map images, index maps, strip indexes, flight line indexes, listings of tech-

*Presented to the Fall Meeting of the Western Association of Map Libraries, La Jolla, California, October 7, 1977.

nical specifications and listings of map parameters to create a cross-index for map searching. When trying to locate a map or remote sensing image, we use the listings first, then go to the cross-indexed imagery file. Most of these listings are of the summary type, containing the parameters of a map or an air photo, not an image of it. These parameters include the coverage area, the extent, suitability information (scale, elevation, focal length, etc.), and a source reference. Then the source reference can be contacted for copies of the indexed material.

We are compiling a number of master record files which will contain information on an entire set of maps. These files will be put on microfiche, giving general locations, general source references and relating the overall parameters. Summary records will describe some of the technical elements, usually on a regional basis. There are also some files being created which will act as unit records and will consist of an image of the actual map or remote sensing image for preview purposes, preliminary to ordering full size reproductions. Key source contacts for such material are the EROS Data Center, NCIC state affiliates and NCIC's four regional offices.

SUITABILITY

There are some basic questions we should all ask when considering the suitability of microfilming: the system nomenclature, the mode (film type), the types of image characteristics which will be shown, such as black-and-ahite or color, and whether it is a written listing or graphic image. In indexing cartographic products, we are interested in the written specifications as well as the graphics. For viewing, most of the written indexes generated by NCIC require 48X magnification, a standard for COM listings. This requirement should be kept in mind when shopping for a multi-lens reader. Other considerations include the source of distribution, availability, costs, updating cycle, the kind of indexing, accessibility to its elements and the total number of records in a system. Closed systems are finite of course, while some of our open ended ones already contain over a million items. As a consequence, under some conditions it is possible to suffer an information glut. This, as well as the other suitability considerations should be kept in mind while we review the characteristics of the systems themselves.

APSRS (COM LISTING)

The Aerial Photography Summary Record System (APSRS) consists of COM listings published in five "strips", or bands, of longitude covering the United States. The four bands covering the conterminous United States are currently available, the band covering Alaska and Hawaii is yet to be published. These catalogs are also being printed in paper copy.

The published bands contain over one million records referencing aerial photography from both public and private sources, including photography both planned and completed. Both COM and paper editions can be viewed in the NCIC public browse file, or purchased for a nominal amount (about $4.00 per strip). The actual imagery is available from those producers listed in the front part of the catalog. The APSRS is updated on a biannual basis.

All of the records in the catalog are done on a 7.5-minute quadrangle basis, within 15-minute bands of latitude and longitude, or by a special county encoding system. Irrespective of the form of input, the information is printed out on a 7.5-minute quadrangle basis. Within each quadrangle, the producer is identified by a number or letter, such as "F" for the U.S. Forest Service.

APSRS (TECHNICAL SPECIFICATIONS CATALOG)

In the APSRS technical specifications catalog, each quadrangle consists of about eight pages of data covering three scale ranges (1:40,000 and larger; 1:40,000 - 1:75,000; and 1:75,000 or smaller, representing low, medium and high altitude photography respectively), and two time ranges (1972 or earlier and 1973 to the present). We would like to add pages recording proposed and in-progress photography as well.

The microfiche output of this data consists of an eye-legible header, which shows the time span covered, scale and geographical location information. Each fiche consists of 15 rows and 18 columns of data, equal to 270 pages of computer printout. Additionally, each fiche has a finding index in its lower right-hand corner.

Each quadrangle cell of data on the microfiche relates the data, the scale, the focal length of the camera, the type of film (black-and-white, color, or infrared), film size, type of photography (vertical, oblique, or, possibly, side looking radar), the amount of cloud cover, and the percentage of the quadrangle covered. The minimum amount of quadrangle cell coverage acceptable in the system is about 20 percent, or about 20 to 30 square miles, although as little as 10 square miles of coverage is indexed in some cases. There is also a "remarks" field. Used by the submitting agency, this may relate such information as the update cycle. Also, there is room for the inclusion of county code information, so that remote sensing imagery may be indexed that way as well.

If two contributing agencies should input similar coverage for the same information cell, only the latest coverage will appear in the catalog. For example, two or three different agencies might have photographed the same cell in the Spring of 1977, within the same scale band. This might happen when NASA flies coastal Southern California, DMA flies military installations along the coast and Mark Hurd flies a contract mission for a coastal county or city. Only the latest data submitted will be shown in the catalog. However, inquiries as to the availability of additional coverage not listed in the catalog can be submitted to NCIC headquarters in Reston, Virginia, on a special form. They will respond with a supplemental listing of all items in the file covering the area in question.

Data is submitted for encoding on two extremely simple manuscript forms. The first is for encoding the latitude and longitude of the southeast corner of the 7.5-minute quadrangle covered. Additional input includes all of the technical specifications, as mentioned above. The second form is for large blocks of photography, on the order of hundreds of square miles. On this form the four corners of a quadrilateral are entered, or a multisided polygon is subdivided into the fewest number of quadrilaterals possible, and these are entered. The computer program then automatically assigns the coverage to the appropriate 7.5-minute quadrangle cells. This form can also be used to enter coverage on a county basis by simple entering the county code and the technical specifications of the photography.

EROS AERIAL/SPACE PHOTOGRAPHY BROWSE FILE

The EROS Data Center Aerial/Space Photography Browse File consists of photo selections generated by aircraft and space platforms maintained by NASA, including Gemini, Apollo, Skylab, Landsat, U-2, and others. This is a unit record system showing actual imagery. It includes black-and-white, color, false color and infrared imagery. The browse file images are on 16mm microfilm cassettes containing about 1,500 frames each. This mode will be changed to a 105mm microfiche system in 1979.

The imagery can be previewed in random access mode on a Recordak Microstar 16mm cassette viewer. This viewer has a 14 inch display screen, about the size of the actual imagery. The image axis can be rotated 360 degrees, useful when viewing space photography where it is often easier to detect significant things from different angles. Ordinarily when viewing aerial photography, shadows should be facing you, but in some cases the imagery can be visually enhanced simply by changing the orientation. This should be kept in mind when shopping for viewing equipment.

Selected frames on the 16mm cassettes can be identified by code numbers printed on paper map indexes. Thus, there is some paper copy associated with this system. Each spot on the index sheet represents an image center. A small dot on one side of the 16mm film frame is a unique code number signal. The viewer locates selected images via this signal, which greatly increases the efficiency of using roll film.

Supplementing the browse cassettes are cassettes containing the technical data, including image coordinates, scene identification number, cloud cover, and path/row identifier. These will also be transferred to COM fiche in the near future.

EROS AERIAL BLOCK PHOTOGRAPHY MICROGRAPHIC ACCESSION AID

A second EROS system is being newly produced, the Aerial Block Photography Micrographic Accession Aid. It contains about 29 photomosaic indexes per microfiche and a second set of microfiche with the technical information for those photomosaics. Where the Aerial/Space Photography Browse File is principally an index to high altitude remote sensing, this is more of a medium and low altitude remote sensing accession aid indexing USGS aerial photography. It will be expanded at a later date to include other agencies.

As reduced as they are, the photomosaics are not tremendously articulate. But, depending on the user's expertise in photo interpretation, they can yield considerable information. The full size paper photomosaics and the individual aerial photographs previewed in this system can be ordered from EROS or NCIC.

As in the APSRS catalog, these microfiche have eye legible headings identifying the region covered. Each "block" is a quadrangle of the International Map of the World (IMW) at 1:1,000,000, such as the Louisville sheet (NJ-16). This block is then broken down into sub-blocks using the USGS 1:250,000 series. The index maps are printed in the upper left-hand corner of the microfiche, or the paper map equivalents can be used as indexes. The index map on the microfiche shows the location of photography. This same fiche also includes the photo characteristics. Since the mosaics are on microfiche, individual fiche can be updated as needed to keep the entire set current, with the superseded fiche forming an historical record of coverage.

This accession aid divides the United States into 53 geographic files, each a 4 degree by 6 degree latitude/longitude span. Each file is then divided into small scale and medium to large scale coverage. The small scale file covers broad areas and is shown on the 1:1,000,000 base maps. The medium and small scale blocks are plotted on the 1:250,000 base maps.

Technical specifications printed on the microfiche include the photo scale, the holding agency, the originating agency (USGS turns over much of its aerial photography to the EROS Data Center), a project identification number and an index identification number, for locating the photography on the microfiche.

MAP AND CHART INFORMATION SYSTEM (MCIS)

The Map and Chart Information System (MCIS) is our attempt to make available an information set on the base maps of the United States. This consists of a complete set of all 7.5-minute and 15-minute quadrangle maps published by the USGS, produced on 35mm roll microfilm. They are in black-and-white, with approximately 500 frames (sheets) per roll. There will be two sets of these, an historic file going back to about 1884 and a current file of newly published quadrangles, issued on an annual basis. Accompanying these reels will be 105mm microfiche listings containing information on the individual quadrangles. Also, in the future, the system is expected to contain a number of subjects other than topographic quadrangle mapping. These would include thematic mapping, special base maps, navigation charts, survey records, economic and administrative maps, digital mapping, historic mapping and related cartographic products such as atlases, globes and literature. It will be available for purchase and its contents will be responsive to user demand.

Presently, the 35mm catalog of topographic quadrangles can only be referenced through the standard paper indexes available from USGS. Since these only show material currently in print, they will have to be annotated to show all of the microfilm coverage where the paper equivalents are now out-of-print.

The MCIS indexes will include such items as quadrangle name, scale, revision date and coordinate location of the southeastern corner, much like the entries in the monthly New Publications of the Geological Survey. There will also be a cross-file listing by latitude and longitude, and a series listing.

MAP SEPARATES INFORMATION SYSTEM (MSIS)

Another microfiche system we have developed responds to the need of the many users who want only a portion of the information appearing on a topographic quadrangle, such as the contours, the Universal Transverse Mercator (UTM) grid overlay, or the hydrography. For these users, we have developed the Map Separates Information System (MSIS). It is an automated catalog which systematically describes map quadrangle reproducable separation materials, including the storage location, geographic position, scale, film base characteristics (scribecoat, mylar, etc.), revision information and latest publication information.

This is a file with limited in-house use, but of interest to anyone who might want to use the reproduction facilities of the USGS to produce a tailored map rather than make overlays or color separations themselves. The catalog lists all information necessary to obtain a given map separation master, one of 10 to 20 different separations possible for a given quadrangle. Using the COM "Charactron" system, we can quickly generate a map order based on the requestor's specific needs. Further information on this system is available through NCIC's Technical Information Section.

BUREAU OF LAND MANAGEMENT LAND AND SURVEY RECORDS

The final microform mapping system uses aperture cards to show land plat information. Aperture cards are 35mm film chips inserted in an otherwise standard data processing card. These are generated by the Bureau of Land Management.

The file is not open for public use, but duplicates can be obtained through the BLM state headquarters offices. Our field survey units use blow-backs from these aperture cards as an information base in creating and updating USGS maps.

These land plat aperture cards are tied into an entire system created by BLM for the preservation and updating of public lands surveys. It consists of cadastral plats, survey field notes, land status and use records, control document indexes, serial registers, tract books and General Land Office (1830-1946) plats. All of these components are further detailed in BLM's Land and Survey Records: Record Preservation & Service Through Through Microfilm, which is available from their Denver Service Center (D500), Building 50, Denver Federal Center, Denver, CO 80225. Each State Office of BLM has information on its land & survey records.

SUMMARY

The following tabular listing represents a summary of the foregoing systems. If there are further questions about them, the staff of the National Cartographic Information Center(s) will be happy to help you get answers to them.

CATALOG SYSTEM	PRINTED INDEX	16mm CASS.	35mm APC	35mm ROLL	105mm COM
Aerial Block Photography Accession Aids					
photomosaic index imagery (USGS b&w)					+
flight line index plots (NASA col, cir, b&w)					+
technical specs. & user instructions					*
block index map (1:250,000/1:1,000,000)	+				
Aerial Photography Flight Line Diagrams Accession Aids - USGS/WMC					
flight line index plots					+
technical specifications (by state)	+				
Aerial Photography Summary Record System (ASPRS)					
5 section graphic catalog	+				+
technical specifications					*
BLM Aerial Photography Accession Aids (cir)					
flight line index plots					+
technical specifications					*
BR Aerial Photography Accession Aids					
photomosaic index imagery					+
technical specifications					*
Cartographic Catalog (CC)					
technical specifications					+
Gazetteer of the National Atlas					*
Historical Reference File (AMS)					
1:250,000 topo. series (color)				+	
cross reference index	+				
Historical Reference File (CE)					
15' topo. series				+	
cross reference index	+				
Historical Reference File (USGS)					
7.5 & 15' topo. series (b&w)				+	
cross reference index	+				
Inventory of Published Maps (alpha sort by state					
technical specifications					*
LANDSAT Accession Aids (ERTS 1 & 2)					
imagery & limited specifications		+			+
technical specs. & user instructions					*
coverage maps	+				+
LANDSAT Computer Compatible Tapes (CCT)					
technical specifications					*
Map and Chart Information System (MCIS)					
example imagery				+	
technical specifications					*
Map Separates Information System					
technical specifications					*
NCIC Map Catalog by Series (alpha sort by state)					
publication & technical specifications					*
NCIC Map Catalog by Coordinates (numeric sort)					
publication & technical specifications					*
Ortho Program Maps (USGS/WMC)					
program map index - 7 western states/ quart.	+				+

CATALOG SYSTEM	PRINTED INDEX	16mm CASS.	35mm ROLL	35mm ApC	105mm COM	+ *
Skylab Accession Aids						
technical specifications					*	
flight line diagrams					?	
imagery		+				
coverage maps	+					
Topographic Program Maps (USGS/WMC)						
program map index - 7 western states/quart.	+				+	

ABBREVIATIONS

AMS Army Map Service (now Defense Mapping Agency Topographic Center)

ApC Aperture card

B&W Black and white

BLM Bureau of Land Management

BR Bureau of Reclamation

CASS Cassette

CE Corps of Engineers, U.S. Army

CIR Color infrared

COL Color

COM Computer Output Microfilm (microfiche)

NCIC National Cartographic Information Center (USGS)

USGS United States Geological Survey

WMC Western Mapping Center (USGS), Menlo Park, California

NOTE: not included in this tabulation are the BLM plat maps on aperture cards. Copies of these are held by NCIC, but, as mentioned in the text, they are available from state BLM offices.

47

A Simple Map Microfilm Program

R. Philip Hoehn, Map Librarian, The Bancroft Library, University of California, Berkeley, CA 94720

The Bancroft Library began a simple, inexpensive and easily started map microfilm program in the Fall of 1977. Since the investigation of the confusing array of products and formats available was the most time consuming part of the process, perhaps a description of the Bancroft system will be of interest to other institutions contemplating a similar project. Much helpful advice and information came from Larry Cruse, Map Librarian at the University of California, San Diego, whose assistance is much appreciated.

Requirements for the map microfilm system were few. First, it had to be one which did not require the purchase of expensive equipment, since the rate of acqusition would, in all likelihood, remain fairly low. The system chosen had to be one which could be set up and maintained by staff with relatively little expertise in microfilm systems. The form of storage had to be compatible with other formats already in the Library. Finally, the records for the microfilms had to be capable of being integrated with records for full size maps.

The project began almost out of necessity, when the Library acquired a reel of film containing about 150 18th and 19th Century maps relating to Mexico. The maps were acquired through a faculty member from originals in a Spanish archive. While the maps were filmed in a technically satisfactory manner in regard to resolution, density and reduction ratio, they were not usable in reel form. The maps were not numbered or otherwise arranged in logical order, and there was no space between frames which might be used to mark them for identification and retrieval. The maps were filmed at a variety of angles (facing up, down, left and right) even for different parts of the same map. Furthermore, the range of geographical areas was too broad to be covered by a reasonable number of subject headings; included were world maps, maps of various parts of the Americas, nautical charts of the Pacific Ocean, and maps of various harbors, forts, and towns. It seemed the only way to make these maps effectively available to the user was either to have photostats prepared from the film, or to cut up the film and put the individual frames into film jackets. The cost of photostats of the approximately 300 frames was a staggering $600. Film jackets to house the same number of frames, on the other hand, was only about $10. The latter option was chosen.

The system of putting film into jackets required only one specialized piece of equipment, a microfiche reader, which Bancroft already owned. Only two non-standard supply items were purchased: the microfiche jackets themselves and lint-free cotton gloves. The microfiche jackets, or "unitized microfilm carriers," are available from a number of sources. Bancroft selected Microseal jackets AS184X1 for 35mm film, and 64105 for 105mm film. The 35mm jackets contain two tracks, or windows, covered back and front by thin plastic. Made of cardboard somewhat thinner than a catalog card, they measure 4 x 6 inches -- the standard size of microfiche. Each track holds two or three frames, depending on size, and is open at the right for the insertion of film. The cost of these jackets is 4.5¢ each; they are sold in 1,000 unit blocks. The 105mm jackets are made of clear plastic, sealed at both top and bottom, but open on both sides. Overall size is 4 3/8 x 5 7/8 inches. They cost 10.1¢ each, also in minimum blocks of 1,000 units. Lint-free cotton gloves can be obtained from most photography stores at about $3 for a package of six pairs. They stretch so that one size will fit almost everyone.

The procedure for inserting film into jackets is simple and can be performed by a student assistant after a few minutes of explanation. The film is handled only when wearing gloves to avoid fingerprinting or scratching it. It is cut or trimmed with ordinary scissors. Placing the film against a white background or on a light table makes identification of frames easier. Single frames can be pushed to the desired position in each track with almost any thin flat object, such as a strip of catalog card stock about 1/2 inch wide. Relatively expensive cutting/inserting machines are available, but are not really necessary for small-scale projects. Commercial firms will cut and insert film for a fee. So far, Bancroft has not used either of these alternatives. Should it become necessary to remove film from the jackets, this can usually be accomplished by gently tapping the open side of the jacket against a table top until the film slides down to the opening. If this fails, the plastic covering the fiche can be torn off and a new jacket prepared. Since the jackets are so inexpensive, this is a feasible alternative.

Numbering or titling can be done on the strip at the top of the 35mm film jackets, either by hand or with a typewriter. This can be done before or after inserting the film. The 105mm jacket has no strip on which to put any information, so Bancroft places a self-adhesive label containing only the fiche call number and the Library's ownership stamp at the top left corner of each jacket. This is a slight drawback, but since 105mm film is large enough so that the title can, in many instances, be read with the naked eye, it is not so great a disadvantage as it might seem.

The fiche in Bancroft have been numbered serially in the same sequence of numbers used for other microfiche in the Library. Map and atlas fiche have a "G" (for the initial letter of LC's map classification system) added to the fiche number, e.g. Microfiche 309G. This was thought to be of potential utility should it ever become necessary to withdraw all these fiche from the collection -- perhaps to locate them in another library unit.

The fiche are cataloged as are other maps or atlases, the only difference being the addition of appropriate physical description notes for the film format, and the occasional absence of a scale statement, if the scale of the original is unknown. Catalog cards are integrated with records for full-size cartographic items.

In addition to the original Mexican map acqusition, other maps and atlases have been obtained in film format. These include a number of fire insurance maps from the Library of Congress and other collections. Multi-sheet atlases, it should be mentioned, are not put into jackets, but are left in reel form, unless they consist of a small number of sheets. Some two dozen 105mm film chips of maps of Vancouver and Victoria were acquired from the Provincial Archives of British Columbia for a very reasonable $1 each. A number of maps of only marginal interest to Bancroft's clients have been filmed on 35mm film by our photo lab and put into jackets. The originals have been sent to libraries where they will be more heavily used, for example, maps of individual mines in Utah were felt to be of more use in a Utah library than they would be in Berkeley. Some other marginally useful maps which were in deteriorated condition have, in lieu of restoration, been filmed and the originals disposed of in a like manner. A number of film negatives of maps in books, which are retained by the library when patrons have photographs made, have been placed in jackets and added to the microfiche collection.

In addition to saving a considerable amount in acqusitions costs over paper copies, other economies have been realized. The approximately 600 map sheets now in microform in the Library would have required one new map case at $250 for storing the same maps in paper form. Acid free folders would have required another $30, and the cost of space would have been an additional, albeit hidden, expense. Even assuming the purchase of the ideal microfiche reader - one with several adjustable magnification ratios - our program has already paid for itself after only a few months of operation.

Several improvements could be made to upgrade the system used by Bancroft, and are being considered. A fiche reader as described above will soon become necessary as our file of fiche grows. A reader-printer for making instant, inexpensive copies at the proper enlargement would be highly desirable. Jackets with the Library's name pre-printed on them would save time. Translucent jackets, instead of opaque cardboard ones, would allow the reproduction of our typed titling information onto duplicate fiche copies. Even in its emergent state, however, the system seems to be working well, and I would strongly urge other map collections to investigate and improve upon it.

The Historical Maps of the City Engineer, City of Los Angeles: a Microfilm Collection *

Bernice Kimball, Chairman of Bureau Historical Committee, Bureau of Engineering, Room 755, City Hall, City of Los Angeles, Los Angeles, CA 90012

I entered the field of historical maps by the back door. I am a cartographer, not a librarian nor an historian. I had been drawing right-of-way maps for the City of Los Angeles. This involves research into maps and documents for each property for street purposes. Because of the research required for many such projects I had become familiar with city maps and records and was therefore assigned the job of locating missing annexation maps which were supposed to exist somewhere in the City Engineer's vault. Were it not for this special assignment, I might still be delineating properties, drawing heavy ink lines over people's front lawns, demolishing those lawns, the plantings and even the front porches, so that other people could drive more safely and swiftly through city streets while the country atmosphere so often desired in the city would necessarily be ruined.

The assignment of finding the missing maps included the making of any maps that could not be uncovered and the correction of any map which did not explicitly delineate the annexed area.

The first step was to dig into the city map files. All maps in the City Engineer's vault are assigned a number. Access to each map relies on its listing in a Division Index as to a general location and in an Alphabetic Index which

*Presented to the Fall Meeting of the Western Association of Map Libraries, La Jolla, California, October 7, 1977.

lists the map under the streets involved or under its title, with but occasional use of cross indexing. There is also a Numerical Map Index which has some value if you know the size of the map and the probable date of its accession.

Since the most frequent need was for maps of streets or of a relatively small location, the Division Index served well. But not all maps were readily available; whenever the area was sizeable, city-wide or greater, or covering more than a few of the divisions, it tended to hide. While such a large scale map was in immediate use, it was known to any of a dozen individuals who were using it. By the time a project was completed, the map number would be engraved in the minds of the many who had worked on the project as well as the location of related but unnumbered maps. As recently as 20 years ago, the engineer in charge of our division required all of his top assistants, some six or seven, to know all active projects going through the office, their present status, the particular problems involved, the councilmen and property owners interested or active in the project, and all important details. It is not suprizing therefore, that maps were located by memory rather than by meticulous indexing.

To complicate the search for basic maps to recreate those annexation maps which could not be found is the fact that engineers consider a map as a tool, not as a separate entity. A map is usually updated to show current conditions or readily discarded as "obsolete". Except for such maps as are published, included in a report, or squirreled away, and for those which represent a static condition such as illuminating an Ordinance or other official action, few maps can be found unchanged. Moreover, the single line entry in the Alphabetic Index confounded identification of useful maps by the brevity and ambiguity of the entry and the practice of dating the map by its date of accession rather than the date of the information it contained.

The next foray for basic information was the History Department of the

Los Angeles Public Library. Here we struck pay-dirt. Not only did the Library have a fine collection of maps of the Los Angeles area, but they were also kind enough to make microfilm copies. Our historical collection was off to a good start.

By car, by bus and by foot, another cartographer and I canvassed our engineering offices in the downtown and outlying district offices, other city departments, engineering departments of the surrounding cities, county offices, regional offices, other libraries, chambers of commerce and any lead to find materials we could borrow. Luck ranged from fantastic (City of Pasadena) to dismal (we cleaned out all the old maps six months ago...six weeks...last week ...we needed the room). We learned not to ask for a "historical" map, but for an "obsolete" map. We learned to ask for an introduction to the "pack rat" of the office, explain our needs and borrow whatever might be useful to the collection, returning originals with microfilm copies for those who had use for them.

The collection of maps of the Los Angeles area now numbers more than 600 maps and seven atlases, ranging from one to over a hundred pages, and includes one map so huge that when it came up in the passenger elevator it had to be cater-cornered while we apologized at each floor for the inconvenience we were causing. This entire collection now fits in a space 3 1/2 x 8 x 12 inches, or less than the space occupied by an unabridged dictionary.

Space requirements were a strong original impetus in the City use of microfilm. The versatility of microfilm has resulted in applications unthought of at the inception of the program. Its acceptance by its users has been complete enough so that most of the older maps can be stored securely, subject to only an occasional recall.

By the time we were bringing in maps for microfilming, there was already a working microfilm system in the Bureau. It did not even occur to me to consider other ways of maintaining files, nor that there were options as to film size and type, or other ways of jacketing the images. The standards already in use by the Bureau were adapted to, and have proven to be workable and good.

Maps are photographed on 35mm unperforated high-contrast film; the reduction is about 1:23 or 1:18. Whereas reels are acceptable for infrequently used material, they have proven cumbersome in use. Most microfilm of maps is jacketed in aperture cards. Microfilm or microfiche taken of books is inserted in plastic jackets; a few are on 35mm, but most lend themselves well to 16mm film. Plastic jackets have not been used for maps since it is advantageous to have single maps readily and separately available. Also, aperture cards and film jackets are not compatible for filing together.

Normally the selection of reduction size and the decision as to the number of frames needed to cover a map adequately has been left to the discretion of the microfilm section. When it appears that a map may be of special value for display, the map is filmed in two ways, once so as to present a good Xerox or other image, and again in a single frame so that it can be reproduced without the problems inherent in splicing two or more sections together. This is the result of slight uncorrected aberrations which even the best lenses, both taking and projecting, can display. If you have ever been around a person engaged in splicing a two- or four-section map together you will know what I mean.

Reproduction of the original silver negatives are easily made by a diazo process. These blue copies can generate a print that is close to the quality of those produced from the silvers; but occasionally the blues are next to useless. The determining factor in their practical use is the careful observance of the standards used in the preparation of the map: line weight, size of lettering, proper space between letters and lines so that no area of the map will blur in the photographing or in the printing process. The blues are sufficiently less expensive that it is worth consideration for use where great volume is involved.

The Historical Map File is receiving a great deal of use. What had been considered to have a rather limited application to city business has proven the opposite in daily use. Now that the information contained therein is readily available, more uses have been found for it. Environmental impact reports have been facilitated and historical land use can be determined far more easily and thoroughly than from examination of other information sources.

The use of borrowed materials has presented few problems; the bulk of the maps are either public records or have long since passed into the public domain. The copying of these materials is generally for immediate consultations. The users are pleased to have a copy they can mark up, lean on, fold, mutilate or spindle. In the few instances where material is restricted or should be attributed to its source, separate agreements have been made with the lenders.

Acqusitions are going on. Some 467 maps in the City Clerk's vault (1850-1900) are being filmed, a dozen or so at a time. A set of obsolete 15-minute USGS quadrangles for the Los Angeles area is in the mill, along with a set of the obsolete 6-minute quadrangles of Los Angeles County, also by USGS. There is active trading going on with libraries. Public records that have gone into obscurity are being ferreted out as time permits.

People sometimes ask me what I am doing. My answer varies from time to time. I was surprised to find myself telling the last inquirer that what I basically was doing was indexing maps and other historical material to make it readily available. That was not what I had thought I was doing. Thinking it over, it became apparent to me that no matter what elegant and interesting finds were made in the line of maps or material, it was of no avail if someone else could not find them. I have sweat blood attempting to find a system whereby "the" map could be found. After trying several systems the present index was created. It was designed to be adaptable to computer cards and to accomodate the information shown on the 20,000 or so maps created before 1930. Even if the numbers never reach the 10,000 or more numbers that make a computer system practical, the use of these cards to present an accurate and up-to-date index will be valuable.

The Historical Index lists maps by date, with an abbreviated title, delineator or engineer's name, and blocks showing the location and specific items of information that have been found to answer the needs of most researchers. By checking the columns it is possible to find a specific kind of map for a specific date.

The machines used for viewing and reproduction are important to the microfilm system. Such machines used by the City of Los Angeles are in the City Engineer's vault and in all offices where there is frequent use of filmed material. Instead of original maps and prints being stored in many locations, only prints of maps or originals that are actually being worked on appear. Almost all maps needed for reference are available on film.

The City Engineer's vault uses several types of reproducing machines. The most satisfactory of these machines are those used for viewing alone. At present, these are Recordak and Microbox readers for 35mm film and a pair of Vantage II's for microfiche.

For viewing and printing on 9 x 12 inch equipment, the vault is now using a 3M "400" reader/printer with interchangeable sets of lenses for use with microfilm or microfiche. This machine is operated by the person requiring the print rather than using the services of the clerks in the vault.

At present the vault rents a Xerox "600" enlarger/printer which produces an 18 x 24 inch print. It can be adjusted to produce an 8 1/2 x 11 inch print as well. This machine gives a long-lived print which, combined with the strict standards of map quality in use throughout the city, results in a fine, serviceable print, though lacking the sharp detail of silver-based products. This machine and its predecessors have been quite satisfactory though not, of course, trouble-free. If the paper furnished is not sufficiently acid-free, the machine will gum up. The drum, depending on the volume of work required, may need replacing sooner than the repairman anticipates. When out-of-service for one reason or another, it may be out-of-order for half a day or more.

The vault also has in its stable an Oce-Elliott reader-printer which can produce either 12 x 18 inch or 18 x 24 inch prints from rolls of paper. It has been furnished with bi-modal paper so that prints can be obtained from either negative or positive film with the simple turn of a switch (if you are blessed with very strong fingers). It produces a fine quality print that has not faded in the few months it has been in use. This machine replaces an Itek machine of similar capabilities (though not adaptable to bi-modal paper) which gave beautiful prints when the chemicals were fresh. The prints, while not permanent, were longer-lived than prints on silver-base paper. They were also among the more expensive printers to operate.

The printing machines in use in my office are:

1. A 3M "200" reader/printer which receives use and is possibly one of the oldest machines still in operation in the city, and one of the most trouble free. It gives a finely-detailed print on silver base paper and prints which can disappear completely in a relatively short time if exposed to sunlight. It will make 18 x 24 inch prints and half-sheet prints of 12 x 18 inches. Magnification is about 23X.

2. Thermo-Fax reader/printer manufactured by 3M. It is another venerable machine and one of the most satisfactory I have encountered, though the screen has become somewhat darkened for viewing. It has additional lenses for varying

magnification of 35mm or microfiche images. It gives a silver-based print that fades abruptly. It is disconcerting to refer to a file folder where the 9 x 12 inch paper has gone entirely white. One of its relatives survived a long time in the City Engineer's vault with some very strange home-made repairs.

3. We have one 3M "400" reader/printer with extra lenses which is only a few years old. We have never quite reached an understanding with this machine. It was almost trouble-free for the first year or so, but now does not always want to print for us. When it works it is very good, giving the well detailed silver print that fades with time or exposure to sunlight. Prints are about 9 x 12 inches

Having been an operator of four IBM alphanumeric printers during war-time, no two of these identical IBM creatures operated in the same manner. Their capabilities, quirks, or failings were all different. The operator learned how to live with the machine -- to coddle it, care for its ills and always check to make sure it was plugged in before expecting it to operate. Likewise, printing machines are temperamental. But with proper care and luck, with hairpins and scotch tape, and with a sympathetic operator, you can get satisfactory performance from many of the various makes available. Because of the heavy volume of work expected from our machines, it could well be that they are more temperamental than those in quieter offices. Good luck in your selection.

APPENDIX 1

How the City's Historical Map System now works:

1. Borrow map by necessary means.
2. Type an index card for each map, map series or atlas. These are not unlike standard library catalog cards, showing the full title, author, publisher, dates, contents, etc. Assign a sequence number. See sample card.
3. Fill out a complete transcript of all data on selected manuscript maps. See sample sheet.
4. Make request for microfilming showing the assigned number and abbreviated title to be punched on the card. Standard request is for six silvers; two are for security and four are for distribution.
5. When both maps and aperture cards are returned they are inspected and checked. Selected data is entered on the Historical Index sheet which is reprinted for limited distribution. Each map is entered briefly on an accession sheet.
6. Distribute aperture cards, revised index sheets and return borrowed material with a set of aperture cards if so desired.

How the maps are being utilized:

1. Direct viewing on reading machines. A complete set of the Historical Maps is available in the City Engineer's vault and in the History Department, Los Angeles Public Library.
2. Xerox prints for reference and marking. Reduced or partial Xerox prints for inclusion in reports or special uses.
3. Machine prints on silver base paper for sharper details.
4. Duplicate films (blues) to substitute for lost silvers or for requested distribution.

5. Photoreproduction to any reasonable size or scale either for display (such as the Stevenson map of 1876 for the Bicentennial) or for publication (as in Neil Harlow's "Official Maps and Surveys of Los Angeles").

6. Photoreproduction on drafting film to permit the creation of a new map.

1909 VIOLE CC
265-2 266-1

CITY AND SUBURBAN STREET MAP of LOS ANGELES
Graphic scale For Photoreproduction

by Felix Viole
Birdseye View Pub. Co, 229 Grosse Bld.

Shows dbl sts, , street railway lines, points of interest, Indian Village, portions of adjacent cities, Ranchos, separate railroad map, City of Watts, Detail map of downtown locating main bldgs. Angels Flight

Index Card - sample

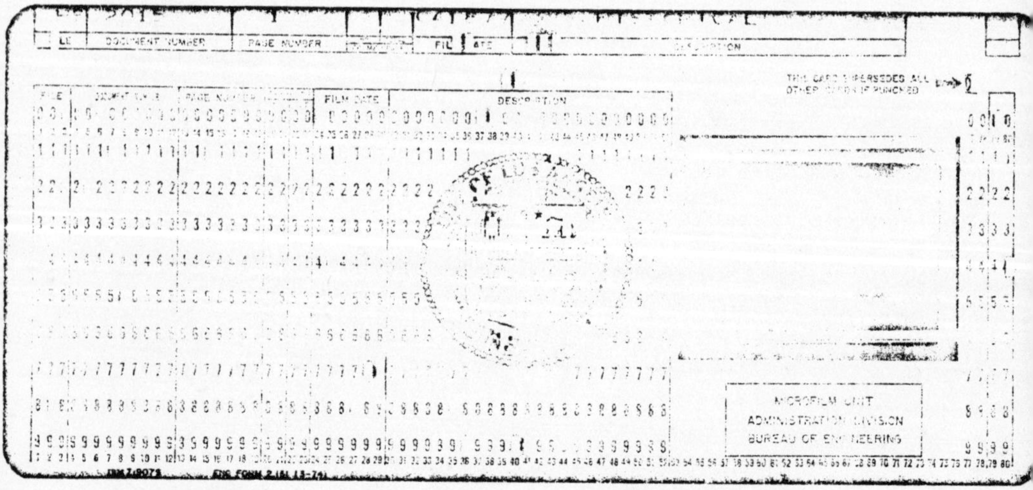

Aperture Card - printed data & card color can be varied

HS2059

No. 57

Size 53 cm 73

TITLE Western Boundary Line of the City of Los Angeles (cont. below)

AUTHOR Frank Lecouvreur, Surveyor

DATE [illegible]

Scale 10 ch = 1" or Graphic Scale ___

Cross Sts from a point 1 League W of the center of the plaza to the SW cor of the City lands

COMMENT ___

Cf Field Book ___

City or Recorded Map ___

MATERIALS

- Ink ✓
- Red Ink ✓
- Pencil
- Tinted
- Pencil adds
- Tracing cloth
- Paper, [illegible]
- Paper, [illegible]
- Paper, mounted
- White
- Brown
- [illegible]
- Photocopy
- Brownline
- [illegible]
- Blue line

CONDITION laminated

- Good ✓
- Fair
- Poor
- Portion missing

CONTENTS

- [illegible] St
- Sgl St
- [illegible]
- Sgl property
- Multiple prop ✓
- Tract
- Lot
- [illegible] ✓
- Bearings ✓
- [illegible] ✓
- Abstract/Title
- [illegible] ✓
- [illegible] Address
- Rancho
- Section ✓
- [illegible]
- Hachures ✓
- [illegible]
- Water courses ✓

D.I. 110, 136, 145, 156, 157

Sample Check Sheet for Manuscript Maps

Sample Sheet from Historical Index

DATE	TITLE	MAKER	DOC. NO.	LOCATION												INFORMATION																				
				COUNTYWIDE	CITYWIDE	OTHER CITIES	CO. WALL SHEET	DOWNTOWN	CENTRAL	HOLLYWOOD	HARBOR	E. VALLEY	W. VALLEY	WEST L.A.	VENICE	DOUBLE LINE ST.	SINGLE LINE ST.	STREET INDEX	OTHER INDEX	RANCHOS/SECT'N	TRACTS/LOTS	DIMENSIONS	OWNERS	ANNEX BORIES	BUILDINGS	RAILROADS	U.G FACILITIES	GEOL-RESOURCES	RELIEF	CONTOURS	ZANJAS	W.S.-RESOURCES	MASTER PLAN	ZONING	POLITICAL SUBD	MAP TYPE
1877	LANDS OF THE L A CITY WATER CO	Hansen/Sol	HS 174						X												X		X			X					X					
1877	COUNTY OF LOS ANGELES (six frames)	Wildy/Stah	HS 255	X													X		X	X	X					X			X							
1877	COUNTY OF LOS ANGELES (one frame)	Wildy/Stah	HS 259	X													X		X	X	X					X			X							
1879	COURSES & DISTANCES OF N BDRY OF CITY LANDS	Hansen	HS 168						X			X					X			X	X	X			X				X							Survey
1881	COUNTY OF LOS ANGELES	Stevenson	HS 258	X												X				X						X		X	X							
1883	SEWERS IN 1883 (Wooden, brick & pipe)	City	4050 Reel					X								X																				
1884	CITY OF LOS ANGELES (one frame)	Stevenson	HS 93		X											X					X		X			X										
1884	CITY OF LOS ANGELES (two frames)	Stevenson	5573		X											X					X		X			X										
1884	ZANJAS, CITY OF LOS ANGELES (based on above)	City	5535		X											X				X	X		X			X					X					
1885	LOTS IN THE FORT HILL TRACT TO BE SOLD	Gillette	HS 175					X								X					X															
1885	MAP OF BLOCKS 75 & 76 HANCOCKS SURVEY (CC 3-79)	City	HS 280						X							X								(Reservoir #5)								X				Survey Notes
1886	CITY SHOWING RAILWAY FRANCHISES	Kellaher	5541 Reel		X											X										X										
1887	CITY (Showing Franchise of Con. Elec. Ry Co)	Rowan/Koeb	5244		X											X					X		X			X					X	X				
1887	ATLAS OF LOS ANGELES (base 1883)	Sanborn	HS 281		X											X		X	X		X	X	X		X						X					
1887	CITY OF PASADENA & VICINITY	Cambell	HS 119			X										X				X	X	X				X			X							
1888	COUNTY OF LOS ANGELES	Rowan	HS 262	X													X			X	X		X			X			X							
1888	OFFICIAL MAP OF LOS ANGELES COUNTY (use above	Rowan	HS 70	X													X			X	X		X			X			X							
1888	ROUTES OF THE CABLE CARS	City	307 & 308		X											X						X				X										
1888	L A, UTAH & ATLANTIC RAILROAD CO -LA to L Bea	LAU A RR	HS 294		X											X				X						X			X	X						
1890	SURVEYS IN SAN PEDRO CHANNEL	Surv Genl	HS 90								X									X		X														

APPENDIX 2

Points to consider in a microfilm system.

I. Whether to film or not.

A. Will there be a real savings in storage space? If so, savings can be redirected to improving archival storage conditions.

B. Cost savings. Indisputable, once equipment is paid for. The City is required to charge $6 each for a map of 9 square feet. Our topographic maps prepared from aerial flights are this size and number 869. The charge would be $5,214. The same information on aperture cards would be charged at just under $61.

C. Original maps can be securely stored and kept free from wear and tear; the acceptance of microfilm and prints from it as substitute is suprisingly high.

II. Whether to invest in cameras and accessory equipment or whether to use a service company which can do microfilming, processing, duplicating, storing, enlarging, and reproducing, or whether resources can be pooled. Keep in mind that the greater the volume, the lower the cost per unit, and that precision photographers do not grow on bushes. Also, it is cheaper to contract for a relatively small volume.

III. Which maps to select. Will all details photograph and reproduce well? Much depends on film chosen, equipment and operators and on the quality of the map itself.

IV. Ask an expert what is really needed in the way of cameras and equipment.

V. Which film or films?

A. Color is delightful, expensive, reproduces poorly, and is evanescent.

B. Color seperation requires at least three exposures and is expensive.

At present, this is the only way of achieving archival permanence for color reproduction.

C. Black and white silver-based film is archival and relatively inexpensive.

1. Negative or positive? Positive looks "right" but negative is easier on the eyes in viewing and gives a sharper print.
2. High resolution or graded resolution? Graded resolution gives a good picture of your dog, but high resolution enhances the sharpness of your maps.
3. 35mm format? We use it as a standard for maps. 16mm is useful for printed matter. Larger film sizes are sufficiently more costly that they are used in the city only where the utmost precision is necessary.

VI. Which reduction ratio or ratios fit your needs?

VII. Determine how many copies are needed and have them done all at once. In our filming, we consider that the first shot costs 25¢ but each additional shot costs only 10¢. You will want at least one copy, perhaps a second for security. It may be desirable for two security copies to be archivally stored for the use of all who have original copies or to furnish copies to other users. Each user may wish to store several security copies.

VIII. Determine how credit should be attributed in such cases as the use of films from scarce materials, or materials still covered by copyright.

IX. Will indexing present a problem? Is it a standardized operation? Can the location of microfilmed materials be indicated on the existing cards or listings?

X. What will serve your printer and viewing needs? Which is essential to getting started and which can be deferred until your operation justifies purchase?

XI. For small amounts of microfilming, the use of any of the better standard

35mm cameras will suffice. If you are purchasing such a camera, investigate some of the new multi-format equipment which can photograph on both 70mm film, as well as medium format, 2 1/4 x 2 1/4 inches. The results from all such cameras will be indistinguishable from those made with professional equipment. Processing will take longer and you must wait to check your results; but the experienced amateur can produce excellent work.

XII. There is no perfect system. Select those elements which most suit you and your needs. You will make mistakes. You will have to deal with the front office. You will make adjustments in your systems, but you can have a compact, comprehensive collection that you think is perfect.

A Discussion of Microfilm Formats and Some Newly Developed Readers *

William R. Lataday, Latady Instruments, Inc., 220 Prospect Street, Hingham, MA 02043

In the late 1950's, the U.S. Army Mobility Command at Fort Belvoir, Virginia, committed funds and engineering and development time to create a completely mobile, truck-mounted system for the reproduction of four-color maps in the field. This was a complete reproduction system capable of producing field usable maps at the rate of 2,000 per hour.

All of the map information was reduced onto 70mm film, either by photographing the original color separations or by photographing a colored map with multiple black and white exposures using the three primary color filters and the fourth negative for the black lines.

The 70mm film was processed in the field, when required, and cut into 70 x 100mm "chips" for placement into an electrostatic printer which projected the images successively through color filters onto the printing paper. Registration holes were made in the chips used in the camera and these holes were placed over pins in the projector to assure that the color information was projected in the correct location.

The map information was usually reduced 8.8 times in the camera to match the fixed enlargement of 8.8 times in the projector. However, the camera was capable of reducing map information from 6 times to 10 times.

---

*Presented to the Fall Meeting of the Western Association of Map Libraries, La Jolla, California, October 7, 1977.

Latady Instruments received the prime contract to design and develop the camera for this mobile system (see Figure 1). Since that first contract, many other cameras, viewing, and projection systems have been designed, developed, and manufactured by us for the U.S. Government. Only those which would be of general interest to the map using community will be discussed in this paper.

A discussion of the 70 x 100mm film chip is germane since it would appear to be an "odd ball" size. In the 1950's, the state-of-the-art for the design and availability of a lens which would give 100 lines per millimeter (l/mm) resolution across the format, the desired resolution, dictated that a format diagonally larger than about four inches was not possible. The 70 x 100mm film chip (actual size 2.5 x 3.34 inches) had a diagonal of 4.17 inches. The next standard film size was 4 x 5 inches or 105mm film cut to either five or six inches long (six inches would be better for the aspect ratio of most maps). If the format was 3.75 x 5.75 inches, the diagonal would then have been 6.86 inches. A lens to cover this format with 100 l/mm resolution was not available and the cost to design and make one not in the budget.

Tests had been made which indicated that the 70 x 100mm film, with 100 l/mm resolution, reduced and blown back at 8.8 times, would produce acceptable map resolution in the printer. There may have been other considerations which determined the final choice of 70mm film, but whatever they were, the 70 x 100mm map system is now a standard for microfilmed maps in many U.S. Government agencies.

A large percentage of all U.S. Air Force maps are reduced to the 70mm chip. Most of the U.S. Geological Survey's maps have been microfilmed onto 70mm chips for inter-governmental distribution. These map files are mostly color transparencies, not color separations as might be inferred from the above discussion.

After the first series of 70mm Micromap Cameras were designed and manufactured, we designed and manufactured a far more sophisticated camera with interchangeable backs for various film widths: from 35mm to 9 1/2 inch and also cut film holders of 70 x 100mm and 8 x 10 inches (see Figure 2). Besides the large movable copy board, the major improvement was the lens (see Figure 3). The lens was computer designed using the latest programs (in 1968) and one of the best computers available at that time in the U.S.A. The design was optimized several times and the Goertz Optical Company was chosen to manufacture this lens for us.

The execution of the design worked well and the resulting lenses were comparable with the computed performance. Only two of these Universal Precision Copy Cameras (Figure 2) were manufactured. One of these two cameras was delivered to the Army Map Service (Defense Mapping Agency Topographic Center) in Washington, D.C. in 1967, and the other to the Air Force's Aeronautical Chart and information center (Defense Mapping Agency Aerospace Center) in 1968. Both cameras have been used constantly (in many cases two shifts per day) to photograph many thousands of maps in color. In this case, since the quantities of copies are few, the original film is used, i.e. reversible film, so that there is no negative for making additional copies. This could be the limiting factor for the availability of these 70mm map copies to libraries or individuals.

Latady Instruments has the only commercially available 70mm Micromap Camera with the 12 inch f/6.3 Precision Copy Lens similar to the one used by the U.S. Government (Figure 3). Plans are being made to photograph all of the coastal NOAA charts in color onto the 70 x 100mm format. By adding a 105mm film back, this Micromap Camera is also capable of photographing maps in high resolution color onto the full fiche format. This size was not chosen for the

NOAA charts because of the additional cost of the larger negative and the lack of a requirement to have a larger format since the 70 x 100mm size gives adequate resolution (about 50 l/mm in color and 160 l/mm in black and white across the total format). Sensitivity of color film is the limiting factor.

Seventy millimeter film has never been considered a standard in the microfilming industry. Only the government has used it extensively for map storage and interdepartmental exchange of information. It has also been used widely in various types of aerial photography.

At a recent NATO meeting dealing with microfilm standards, the 70mm format was retained as an international military standard. There are some indications that it may also be used soon as a substitute for 35mm in aperture cards where more information is required. The standard aperture card will accept the 70mm film width, only the window must be enlarged. But such a change to 70mm is not expected to be universal.

In 1975, the U.S. Army Mobility Command, acting on behalf of the U.S. Air Force, contracted with us to design and develop a 35mm microfilm storage and film reader for the use of airmen who might be downed in unfamiliar territory. This unit (Figure 4) will hold up to 90 maps on "thin" film, i.e. 2 mil film, reduced to a format size of of 1 1/4 x 1 3/4 inches. The unit contains a compass and batteries for a light for night use. It is both water and dust resistant for field use. Unfortunately, it is too expensive to produce for commercial sale.

A simpler model of the above 35mm (MICROVIEW) storage and film reader has been designed and a prototype manufactured. This unit is about half the size (1 1/2 x 2 x 3 1/2 inches) and will weigh about five ounces. It will hold 15 maps on standard 35mm black and white or color film. This unit is planned for production sometime in 1978 (see Figure 5).

To utilize the 70 x 100mm film chips, another very simple viewer has been designed and a prototype constructed. It will hold a single film chip, either black and white or color, which has been laminated between thin sheets of mylar for protection (see Figure 6). A separate carrying case or file case will hold additional maps, charts or other information for storage. These cases will be of a size suitable for field transportation or office storage. Another feature of the 70mm separate film chip will be the heading or map/chart number which is photographed in large bold type on each chip. This heading will be easily read without the use of a lens. These numbers will also be exposed in each file case.

The last of the family of hand-held microfilm readers is a larger unit which will accomodate microfiche. A viewing lens which moves along two axes will cover all of the images on a standard microfiche. The design has been completed and a prototype manufactured (see Figure 7). It is only slightly larger than the fiche and is one inch thick. This unit can be used in the field or office for viewing standard fiche or maps and charts reduced to any size on the fiche. The unit will also accept two 70 x 100mm chips (useful for those wishing to follow a continuity from one chart to another, or for the comparison of two maps).

The 105mm microfiche viewer is also planned for production in 1978. Both the 70mm and 105mm viewers will accept lenses of several different magnifications, i.e. 10x, 15x, and 22x.

The final unit on the drawing board is a waterproof, shockproof microfilm/microfiche projector. This unit is designed for field use where water and dirt may be present in sufficient quantities to make a standard viewer inoperative. In addition, this viewer has three lenses for projection in magnifications of 6x, 12x, and 18x. The lower magnifications are used to project a large area of a map or chart while the higher magnifications would show more detail on the map or chart and the highest magnification is to project 24x microfiche. It is designed to accept the standard 4 x 6 inch fiche, two 70 x 100mm chips, and any smaller negatives or positives. Also, with the addition of an adaptor, it will hold and project 16mm or 35mm roll microfilm at the same magnifications as above. Limited quantities of these projectors will be available in 1978.

Figure 1

70MM LATADY CAMERA

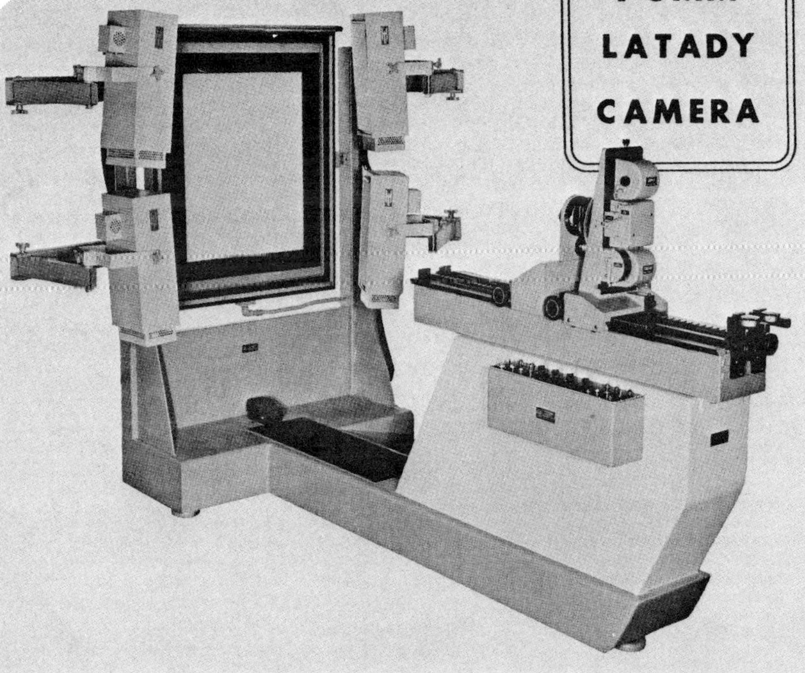

This 70mm Camera reproduces any opaque or transparent map, photograph, line drawing or artwork up to 32" by 40". Material is reproduced onto a frame or "chip" of 70mm roll film. This "chip" serves as a data storage device.

TECHNICAL DATA:

Copyboard size	32" x 40"
Method of holding copy	Vacuum with Mylar cover
Lighting—rear	Dimmable fluorescent
Lighting—front	4 Pulse Xenon lamps
Lens	8¼" Microtar
Resolutions	100 lines/mm
Distortion	Less than 5 microns
Scale (down)	From 6:1 to 10:1
Shutter	Shutter speeds 1/100, 1/50, 1/25, 1/10, 1/5, 1/2, 1 to 111 seconds in .01 second increments
Weight	Approximately 2000 lbs.
Size	Approximately 10' long, by 6' high by 4' wide
Vibration isolation	Camera is mounted on three leveling shock mounts
Delivery	60 to 90 days

- Micro Circuit
- Micro Filming
- Map Copying
- Precise Data Storage
- Digital Stepping Grids
- Storage of Precise Drawings
- Plain Grids
- Color Separations
- Laminate Copying

LATADY INSTRUMENTS, INC.
220 PROSPECT STREET
HINGHAM, MASS. 02043

Figure 1a

Operation—The 70mm Micromap Camera combines the elements of a copy camera and a 70mm roll film camera in one unit. It can be operated entirely by manual controls or semi-automatically from a centrally located control panel. The panel is in front of the camera where the operator would normally stand. The front and back lights as well as the punches, the timer and the opening and closing of the shutter are controlled from this point.

Copyboard—Holds up to 32" by 40" *opaque* material or 24" by 30" *trans-illuminated* material. Copy is held by a .005 inch sheet of Mylar which is pressed against the glass copyboard by a vacuum system controlled by a foot switch.

Lighting—Opaque copy is illuminated by four Pulse Xenon lights in front of the copyboard. These lights, mounted on jointed arms, can be swung towards or away from the glass surface of the copyboard to vary the coverage or intensity of the light.

Transparent or semi-transparent copy is illuminated by fourteen green fluorescent tubes (peak at 5300 ang.) placed on a radius to compensate for the COS[4] fall-off of the lens. The fluorescent tubes are dimmable over an intensity ratio of 4:1.

Scale—From 6:1 to 10:1 is controlled by a step and dial system. Scale is achieved by placing the front and back camera supports at predetermined positions; the one inch steps and one inch range, .001 inch reading dials, provide accurate positioning of both the lens and camera back for any magnification.

Back Support—Flat casting, machined parallel to the copyboard to ± .0005 inches. On this back are mounted the supply cassette, the camera back, the take-up cassette and the gear box for single frame film advance. The camera back holds the pressure plate, the fil punches and the film cutter. It is permanently attache to the back support, but swings away to allow camera loading. An interlock prevents damage to the punches should they be "in" when the camera back is swung open.

Cassettes—Supply and Take-up cassettes each hold 100 feet of either color or black and white film. They are light-tight: film with a leader can be loaded in daylight and non-leader film is safe after dark room loading. The cassettes are independent: several supply and Take-up Cassettes can be used interchangeably with different film types.

Lens and Shutter—This 8.25" lens was designed and manufactured by the Goerz Optical Company especially for this camera. The lens resolves 100 lines across the total field of 2-9/16 by 3-7/16 with less than five microns distortion.

The shutter and shutter solenoid are shock mounted on vibration isolators so that the shutter action does not affect lens resolution. Shutter speeds of from 1/100 to 111 seconds are obtainable.

Vacuum System—Pump produces vacuum needed for the copyboard and collects punched film bits. The vacuum pump is controlled by a foot switch at the base of the copyboard.

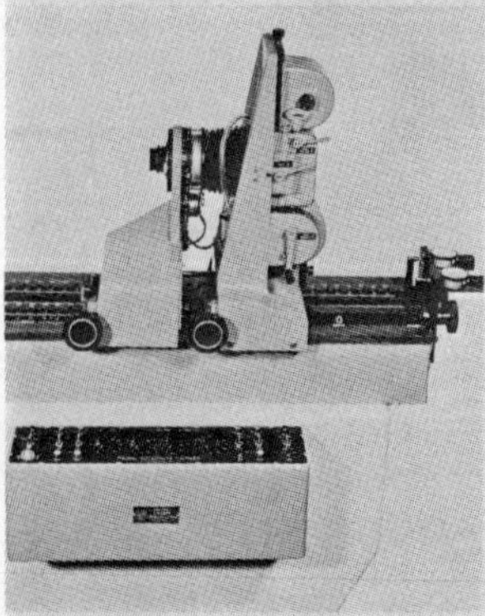

Camera & Control Panel

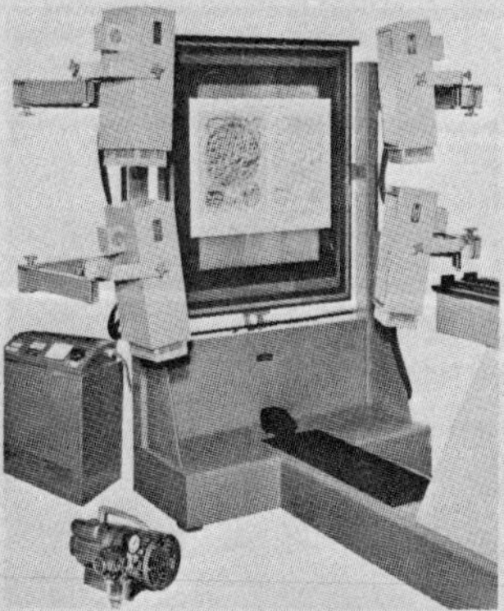

Easel (Copy-Board) Xenon Lights & Pump

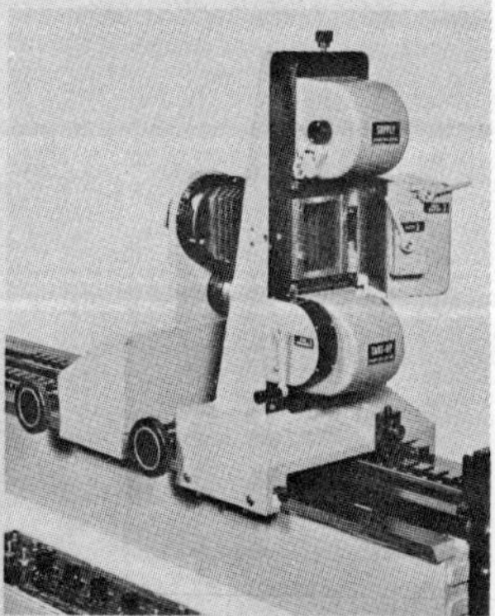

Back of Camera Showing Cassettes

LATADY INSTRUMENTS, INC.
220 PROSPECT STREET
HINGHAM, MASS. 02043

Figure 2

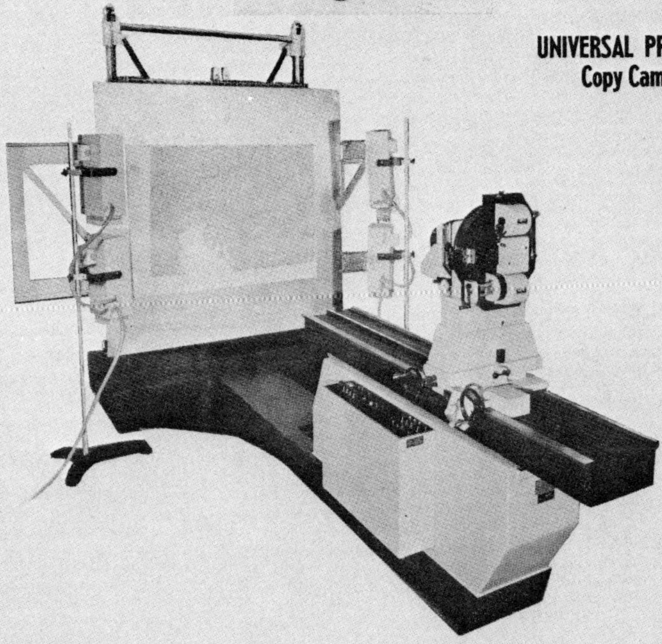

UNIVERSAL PRECISION
Copy Camera

This Universal Copy Camera is designed to copy maps, line drawings, color separations or mosaics up to 75" by 75". The step and repeat capability of the copyboard and the camera head make possible the reproduction of selected sections of any area within the 75 by 75 inch area.

TECHNICAL DATA:

Copyboard size	75" by 75" — detachable
Copyboard movement	±21" in "X" and "Y"
Method of holding copy	zoned vacuum
Lighting	4—1500 watt pulsed xenon tubes
Lens	12" f/6.3 Latady Precision
Resolution	160 ℓ/mm on axis 140 ℓ/mm 9 inch diagonal 80 ℓ/mm 12 inch diagonal
Distortion	approximately 15 microns
Scale (12" lens)	4X to 10X reduction
Shutter speeds	1/100 to 111 seconds
Weight	approximately 4000 pounds
Size	10' high by 10'6" wide by 17' lg.
Film backs available	9" x 9" cut film 70 x 100mm roll film 70 x 100mm prepunched cut film

Other film backs available upon request.

Precision

This copy camera is built with the utmost precision throughout. The copyboard, the camera head, the scale and focus are all setable to ±.0005". The lens has a resolution exceeding all the known color films and many of the black and white films presently available.

LATADY INSTRUMENTS, INC.
220 PROSPECT STREET
HINGHAM, MASS. 02043

Figure 2a

Design Features—The Universal Copy Camera is designed to accommodate individual backs to make high precision "chips" or continuous film strips of a size from 70mm up to 9 by 9 inches. Each film back is located on a combination of cylinders and flats so that registration from one back to another is better than .0005 inch. The focal plane of each back is set to better than .001 inch. At present, three backs are available, ie. 9½ by 9½ inch cut film, 70mm roll film and 70 by 100mm cut film. The design allows for additional backs such as 105mm roll film, for microfiche 5 inch roll film, 4 by 5 cut film or any other size up to 9 x 9 inch.

Lens—The 12 inch f/6.3 Latady Precision Copy Lens is a break-through for resolution across a wide field working at these finite conjugates. This lens was designed by this company using high speed digital computers. The resolution quoted on the reverse side will meet or exceed the resolution requirements for all of the flight navigational aid systems presently under design.

Copyboard—The 75 by 75 inch copyboard is made of magnesium so that two men can easily lift it off the stainless steel mounting pins (it weighs 105 lbs). When returned to its place, it seats to a position, ±.001, thereby, remaining collimated to the lens axis. The copyboard is removable so that aerial mosaics can be mounted on its surface while it is horizontal. Vacuum may be applied while in this position and the board replaced on the camera. The copyboard has four areas which can be evacuated separately. Each zone is ¼ of the total area or 37½ by 37½ inches. The position of the copyboard is shown in both .001 inches and .01 mm by means of Trav-a-dials in both X and Y.

Lighting—Four standard, 1500 watt Pulsed Xenon Lights are provided on two separate stands. These lights are controlled from the main control panel.

Scale—The 12" focal length f/6.3 Latady Precision Copy Lens gives a scale reduction from 4:1 to 10:1. Other reductions can be achieved by substituting other lenses, however, a careful selection must be made to give satisfactory resolution for the format chosen.

Available Backs—9" X 9"—The camera has a 9" X 9" (9½" X 9½" film) cut film back with 10 cut film holders. Each cut film holder has a vacuum platen which holds the film against the film plane during exposure. An ingenious method allows the use of a dark slide to protect the film when the film is not pressed against the focal plane.

70mm Roll Film—The standard 70mm roll film back with the punches and cutter used so successfully on the Latady 70mm Chip Camera has been adapted to form one of the backs available on this camera. This back may be taken off at any time (without exposing the film) and placed back on with the same ±.0005 inch registration as the 9 X 9 back.

70 X 100mm "Chip"—Our standard 70 by 100 cut film "chip" holder with the two square holes on one side has also been adapted to fit on this camera. This back incorporates an automatic pin registration feature which locates the film in relation to the copy to ±.001 inch.

Projectors—Each film format area has a corresponding projector back which projects the corners and a center cross onto the copyboard. This will show the area covered by the different formats at varying scales and allow the operator to select the portion of the map he wishes to copy.

Controls—The main control panel controls all of the lights, vacuum pumps (2), the copyboard X and Y movements, the camera azimuth movement as well as the timer, punches and shutter operation. The copyboard X and Y movement may also be controlled from a remote pendant box.

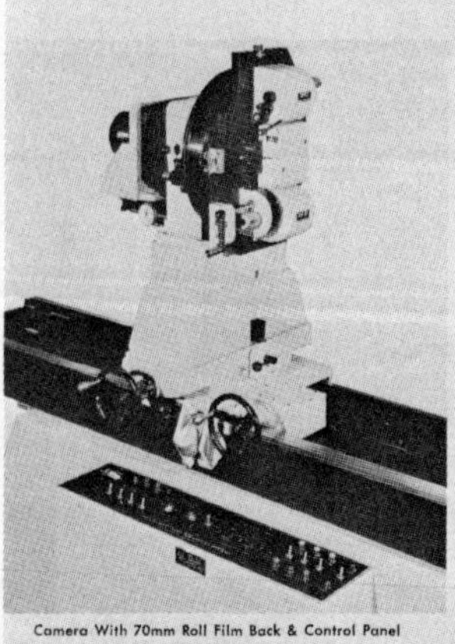

Camera With 70mm Roll Film Back & Control Panel

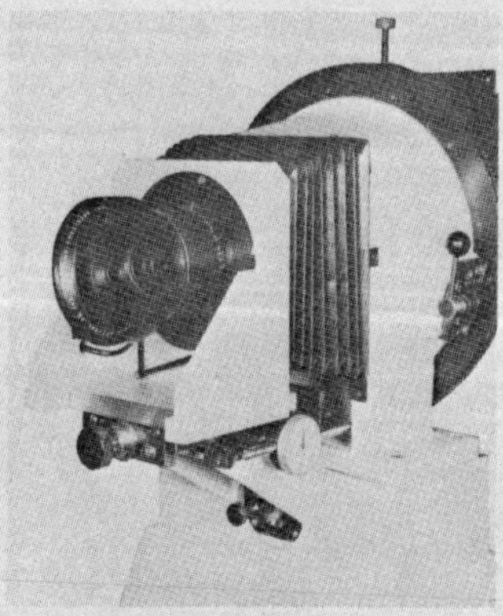

12" f/6.3 Latady Precision Lens

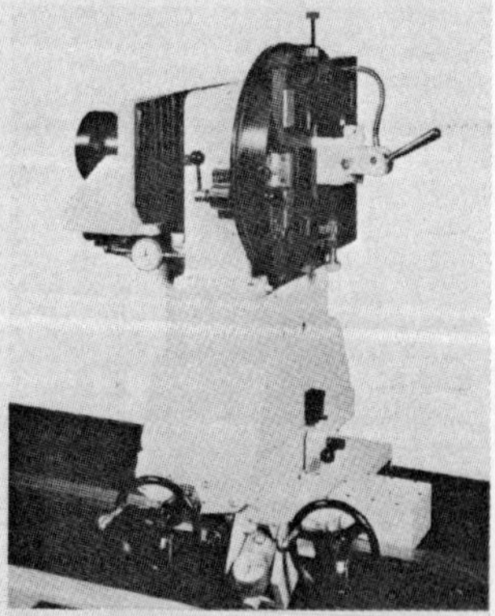

Camera With 9½" x 9½" Cut Film Back

LATADY INSTRUMENTS, INC.
220 PROSPECT STREET
HINGHAM, MASS. 02043

Figure 3: Latady Precision Copy Lens

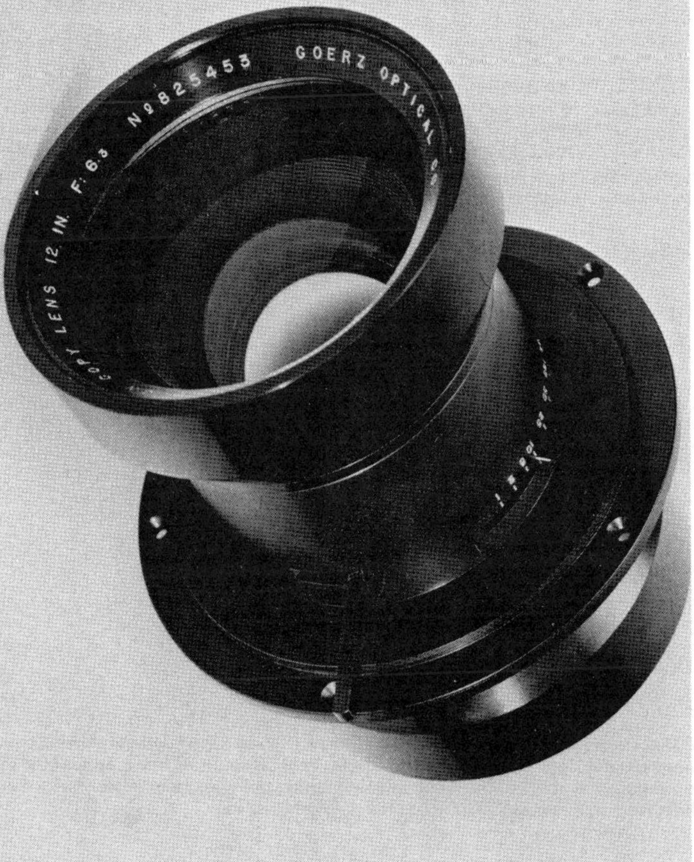

Figure 4

LATADY™ MICROVIEW

80

PAT. PENDING

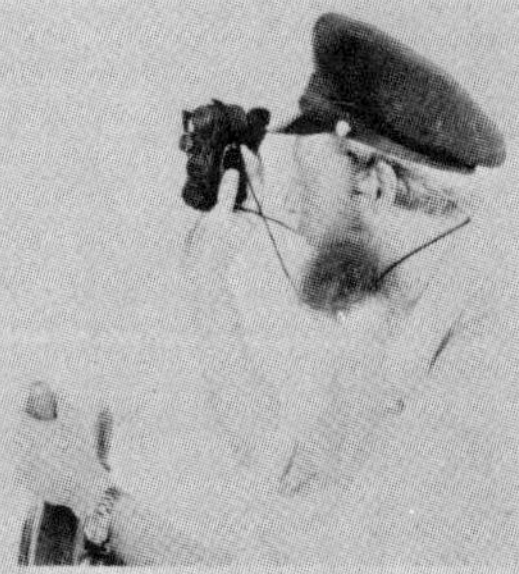

INTRODUCING

A QUALITY LIGHTWEIGHT AND COMPACT STORAGE/FILM READER

The Latady Microview provides a means of storage and viewing microfilmed maps, charts or printed material in a lightweight, water-resistant container. It also provides a means of finding your position and plotting a course. A high resolution 15X lens moves in X and Y to cover the 1 1/4 X 1 3/4 inch image. This lens contains a reticule for sighting, and a reference azimuth ring for determining a heading. A luminous compass swings out when needed. Accessories include a light for night use and a vinyl carrying case.

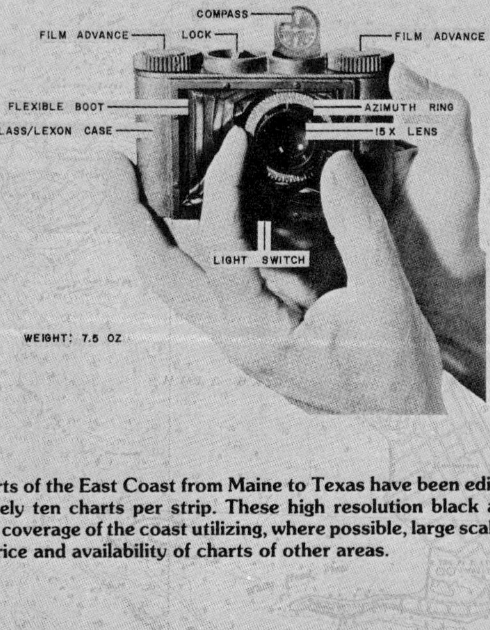

NOAA charts of the East Coast from Maine to Texas have been edited into chart strips of approximately ten charts per strip. These high resolution black and white films give a continuous coverage of the coast utilizing, where possible, large scale charts. See attached sheet for price and availability of charts of other areas.

LATADY INSTRUMENTS, INC. ■ P.O. BOX 39 ■ ACCORD, MASSACHUSETTS 02018

Figure 4a

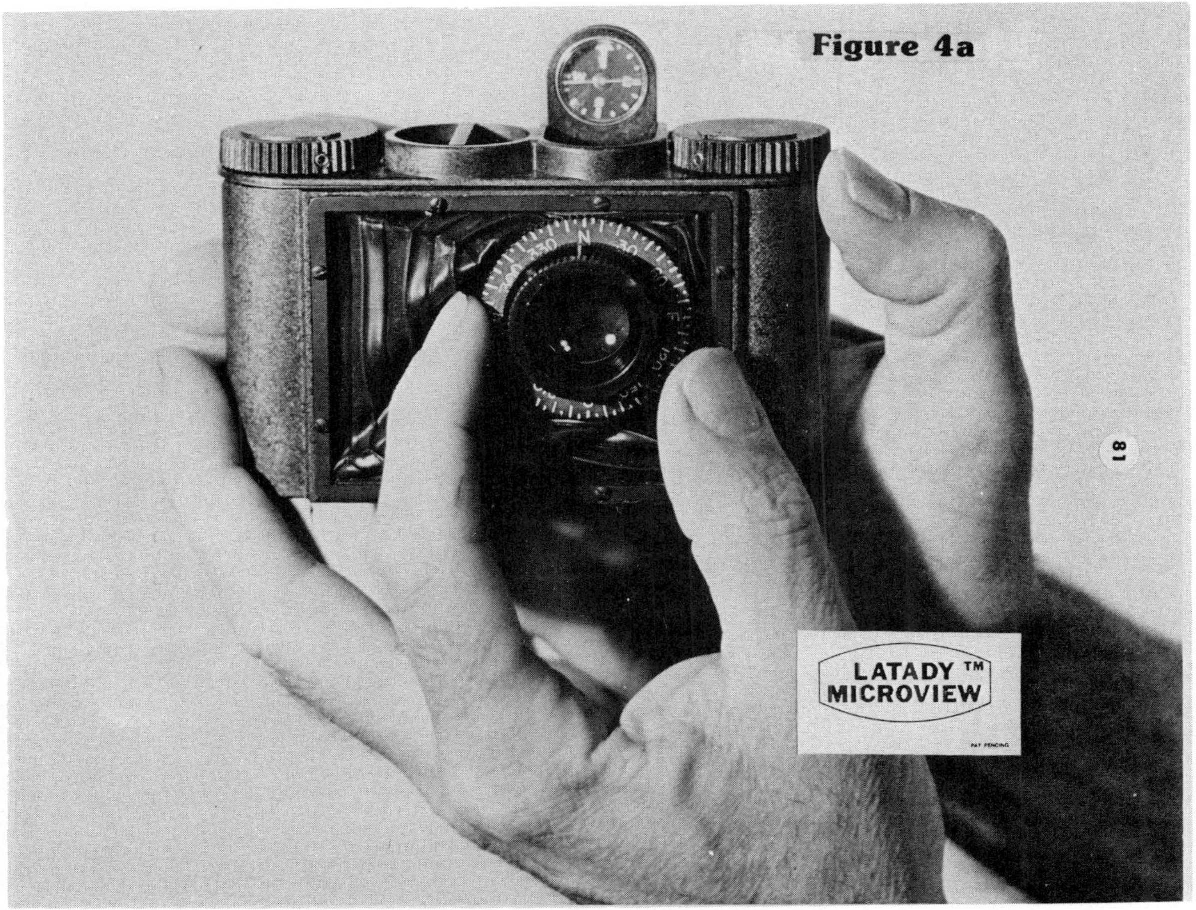

Figure 5: Latady MiNimap Viewer

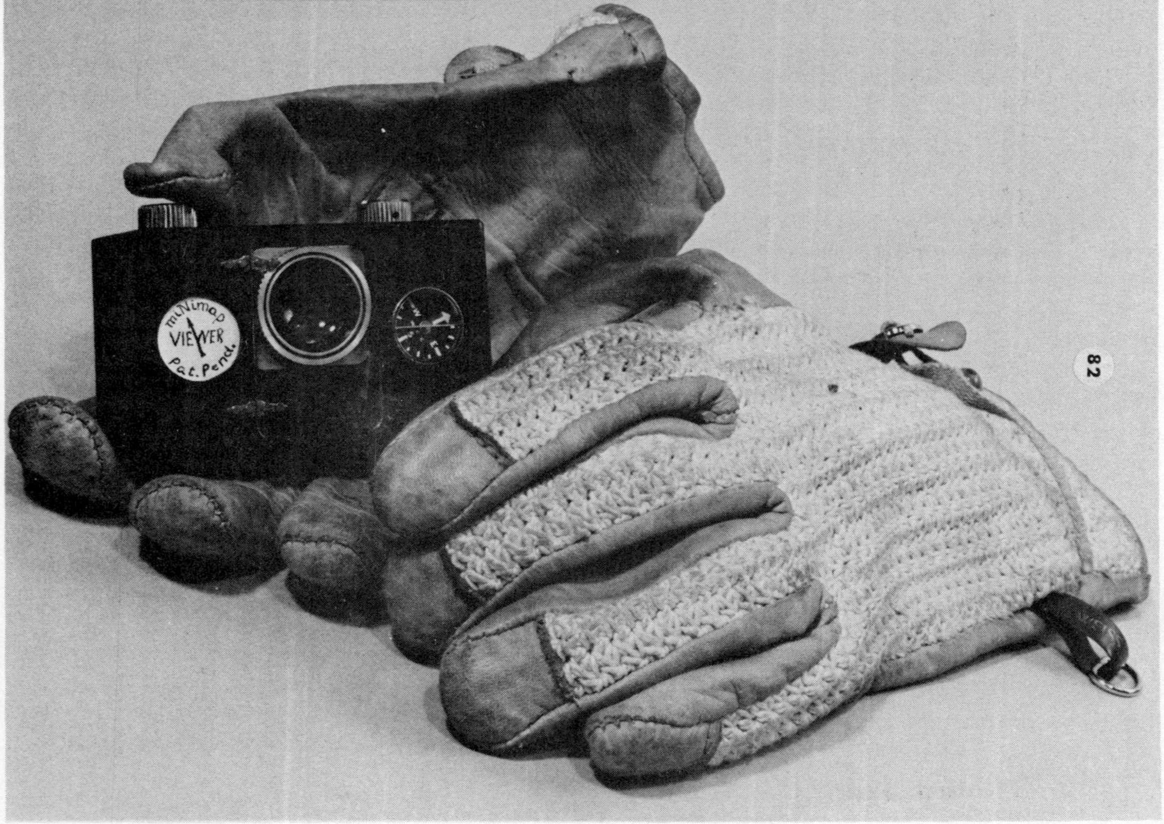

Figure 6: Latady Viewer for 70 x 100mm film chips

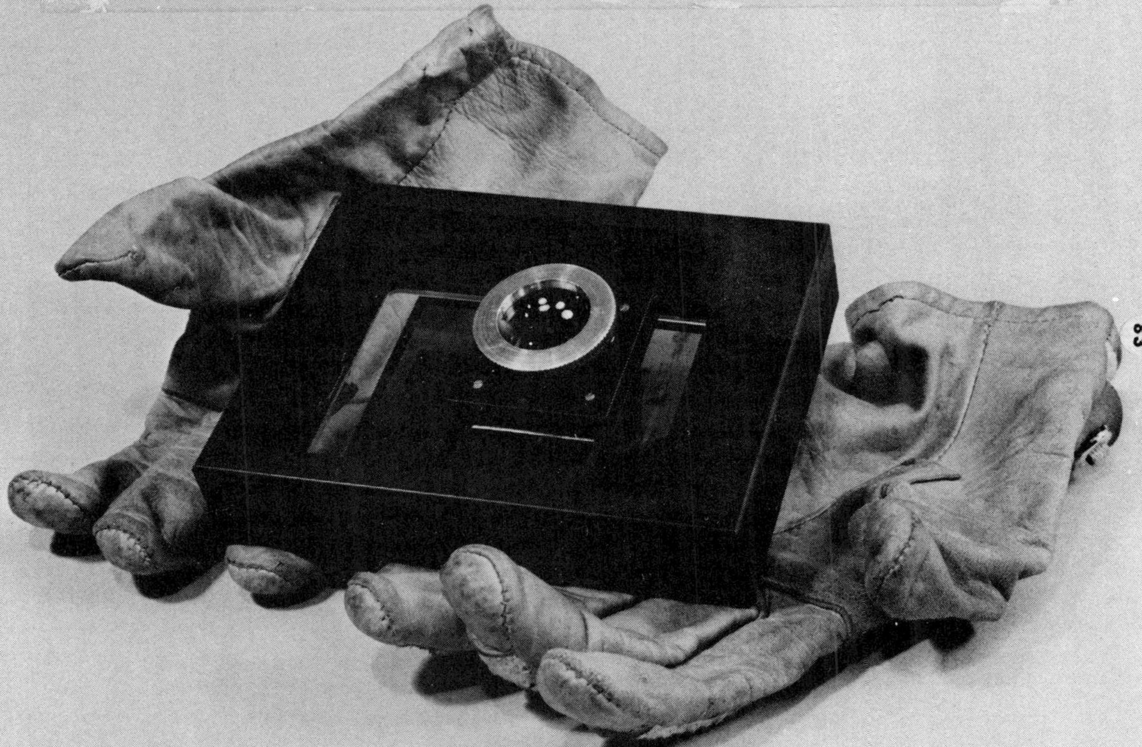

Figure 7: Latady 105mm microfiche viewer

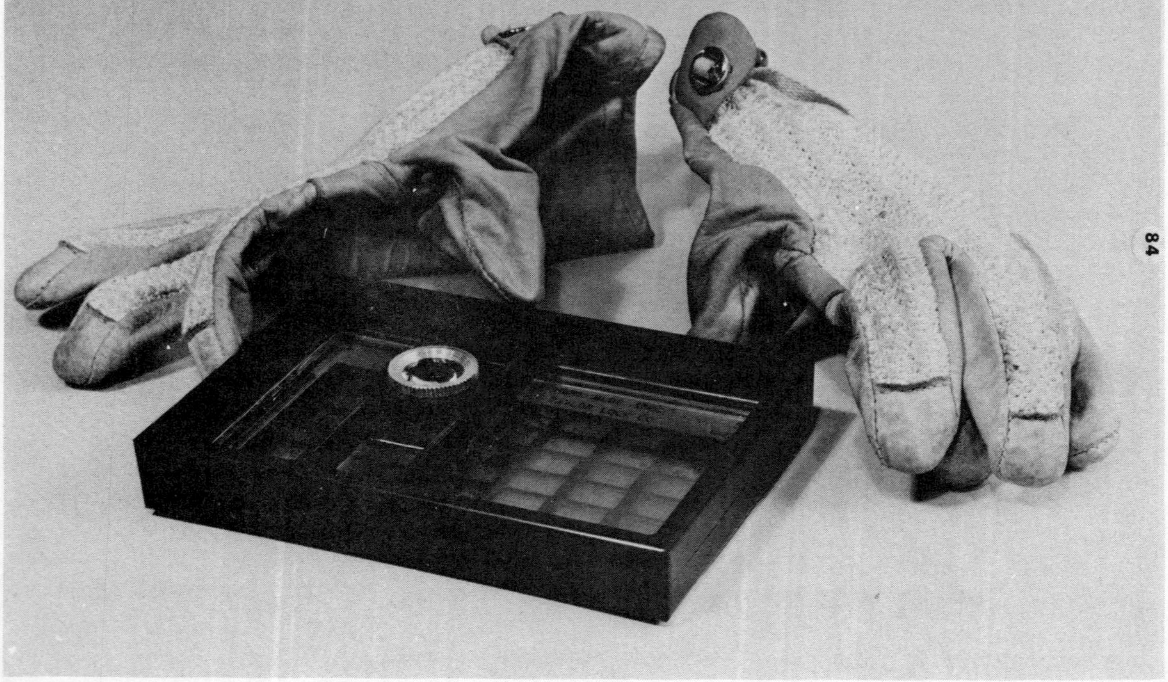

The Microfilm Programme of the National Map Collection *

Gilles Langelier, Head, External and Internal Services, National Map Collection, Public Archives of Canada, 395 Wellington Street, Ottawa, Ontario K1A 0N3

INTRODUCTION

The National Map Collection began microfilming its collection of maps, plans, and architectural drawings in the summer of 1977. Two years previous, after studying the various systems in operation for maps, it was decided to use the 105mm format. It was proven to our satisfaction at that time that this format was superior to the others. To summarize, we found the major advantages to be its low reduction ratio, high resolution, and its ability to reproduce large maps on one frame. In addition, the photoduplication services in the Public Archives could be entirely based on the 105mm microfilm negative.

GOALS OF THE PROGRAMME

With this programme we are trying to solve our major conservation problem which is the constant handling of the original documents by staff members and researchers. A major conservation study several years ago indicated that our collection was deteriorating at an alarming rate.

The microfilm programme will also ensure a better diffusion of our cartographical documentation. We have already received comments from map collections expressing interest in the purchase of diazo copies of maps relating to their regions. Our programme has also attracted a great deal of interest from other government departments which use our collection extensively. We are planning to

*Presented to the Fall Meeting of the Western Association of Map Libraries, La Jolla, California, October 7, 1977. Updated to November 1978.

build a microfilm room where researchers could work not only during business hours but also overnight. The textual records of the Public Archives have always been available to the research public on a 24-hour, 365 days a year basis. However, the nature of cartographic materials prevented us from providing this service previously.

SCALE OF THE PROGRAMME

Our programme is not limited to manuscript maps and drawings. It includes printed maps, transcripts, photocopies, blueprints, etc. At one point we did consider limiting the programme to manuscript and printed maps, but since transcripts, photocopies and blueprints are interfiled with the manuscript and printed maps, it would have been too time consuming to divide them into two groups. Also, and more important, researchers would not have had complete access on microfilm to a specific group.

Although we are microfilming everything, we have established priorites. We have started with the maps that are fully catalogued (approximately 60,000). These are the maps described in the 16 volumes of our catalogue reproduced by G.K. Hall & Company last year. Within this group, first we are filming the maps filed in small folders (20 x 30 inches). Next, we will film those oversized maps which suffer most from handling.

After completing the filming of the catalogued maps, we will microfilm those maps and plans transferred from government departments. It is anticipated that this will take at least several years.

Eventually we hope to film topographic, geologic, or other types of series maps. However, this will depend on developments in the coloured film available.

NUMBER OF COPIES

Two exposures of each map are being taken. The first copy is being kept

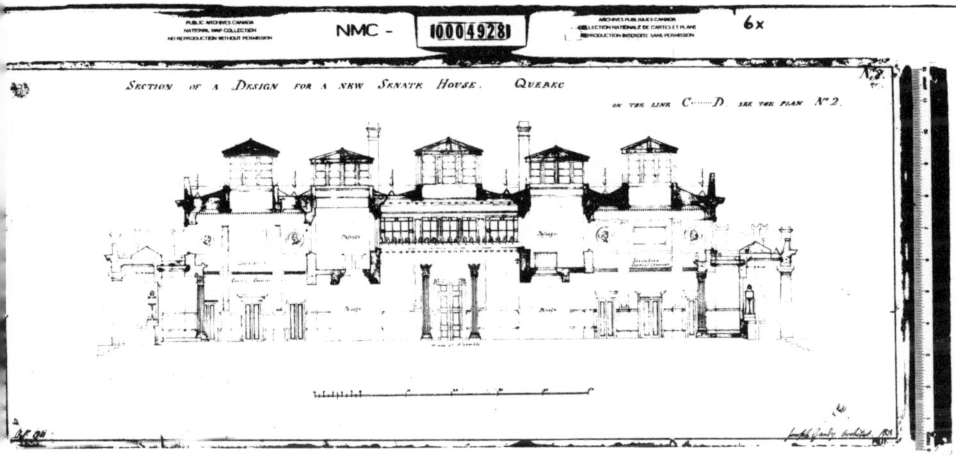

Illustration 1 Architectural Drawing

Illustration 2 Mappemonde ... Couel ... 1765

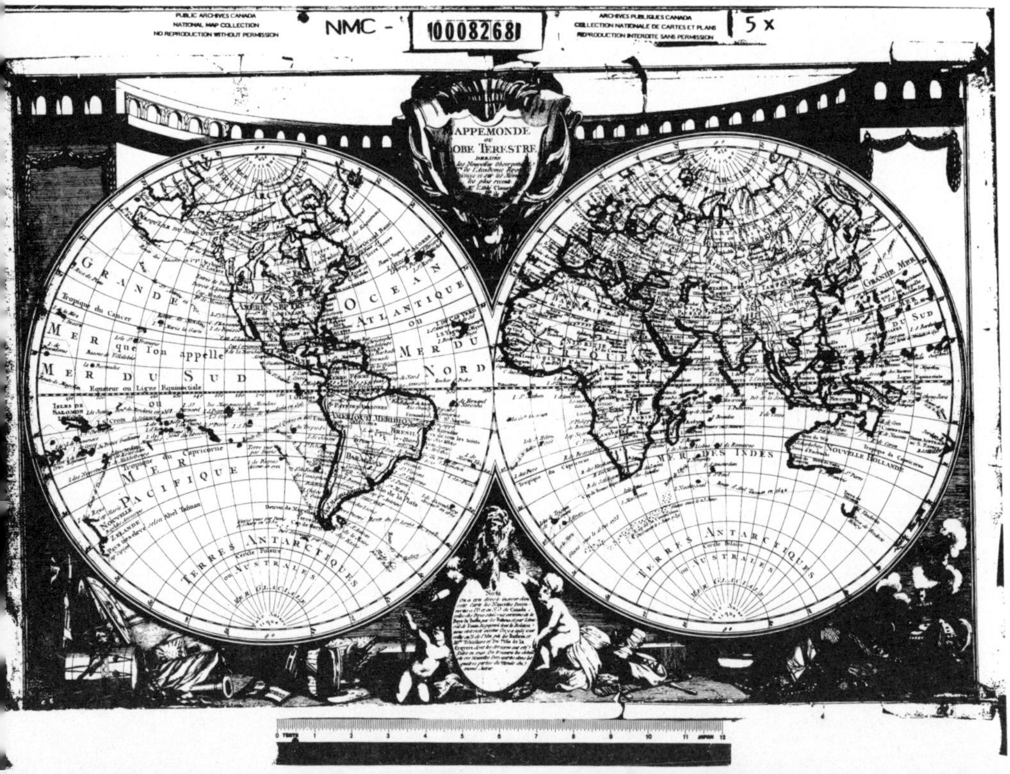

as an archival or security copy. It will be stored in the best environmental condition in a building other than the Public Archives. We have been considering having a cold storage room built to store not only our 105mm masters, but also the numerous microfilm masters already made for the various divisions within the department.

One diazo copy will be available in the reference room of the National Map Collection. This will be filed by call number. In addition, a contact print will appear on the catalogue card, linking the microfilm programme to our automation programme. This is still in the planning stages.

SERVICES OFFERED FROM THE 105mm PROGRAMME

As previously mentioned, the 105mm microfilm programme will drastically change our photoduplication services. At the moment, we rely on the photostat process to produce copies for our researchers. However, the machine in use is no longer mechanically reliable and it is becoming increasingly difficult to obtain the proper paper for the process.

The photostat will be replaced by photographic enlargements from the 105mm negative. These will be available to a maximum size of 40 x 60 inches, and will be made from one of the master copies.

In addition, we plan to have a 105mm reader/printer in the National Map Collection. This will enable us to provide working copies the same day as the request. Contact prints will also be available and it will be possible for researchers to order diazo duplicates. Institutions in particular will be interested in this latter service. Perhaps I should state at this time that we will not reproduce for researchers those items the originals of which are held by other collections or which are copyrighted or have donor restrictions.

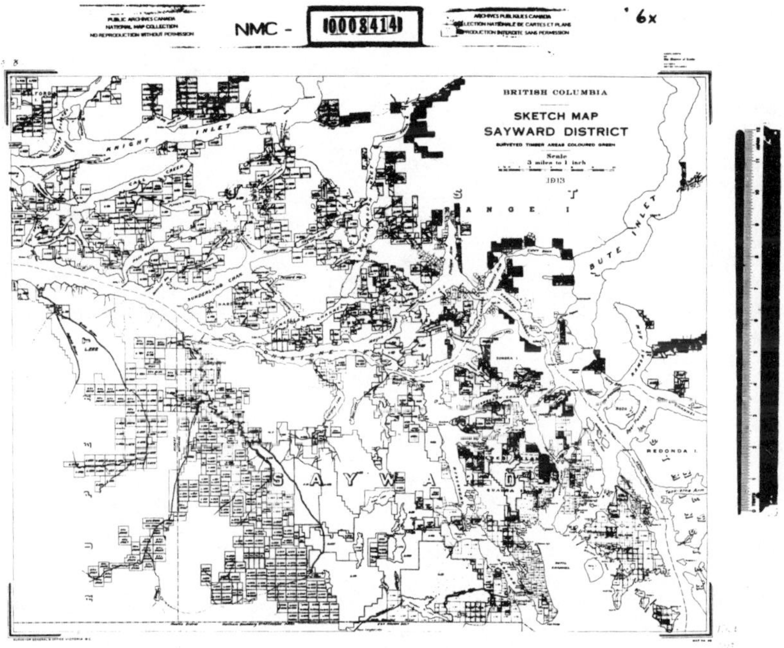

Illustration 3 Sayward District, B.C., 1913

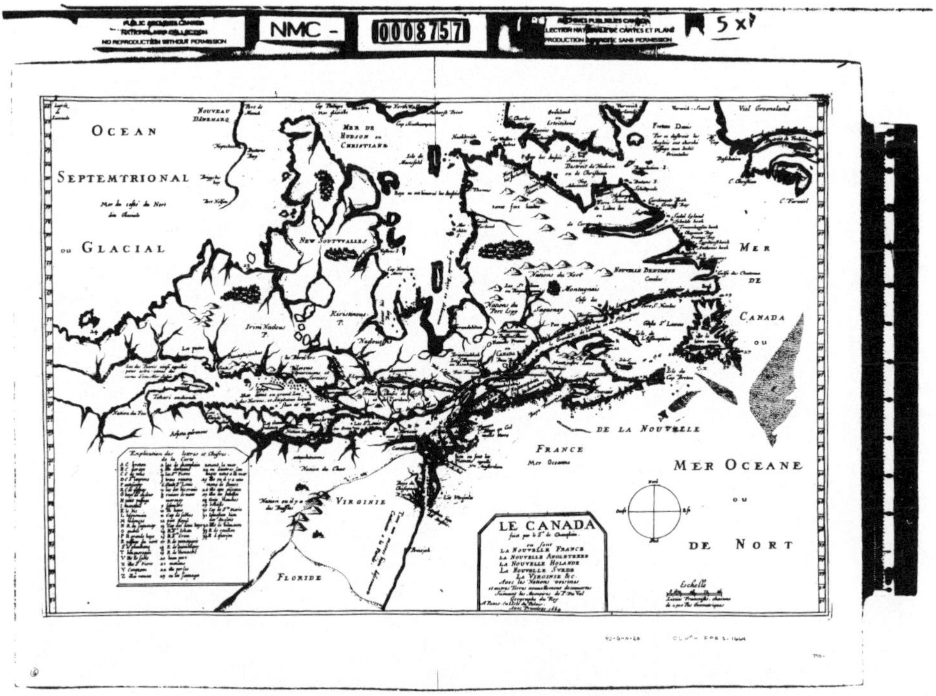

Illustration 4 Le Canada ... Duval ... 1664

STORAGE ENVELOPES

We are using three different types of envelopes for the various microfilm produced:

Diazo: Transparent sleeves with outside headings.

1st Master (Archival Copy): We are using Kodak storage envelopes for processed film. We seal them on the outside with special non-acidic tape (Filmplast P.).

2nd Master (Stored in our Reprography Section): We will be using a custom-made envelope of acid-free paper.

DENSITY

We are trying to obtain the best density for this type of material, which is between .9 and 1.2. Any film which is lower or higher must be redone. During the first month we had to microfilm several maps twice. The main problem we have found is how to obtain a good density on photocopies. Printed maps are the easiest, and in most cases, we are successful the first time. We also had trouble with manuscript maps during the first weeks but we are now refilming only 5%.

RESOLUTION

Our main concern right now is related to the results we are obtaining from our Keuffel and Esser (K+E) camera. I must say that we are just within the limits of the proposed National Microfilm Association (NMA) standards although the technicians from K+E were able to obtain excellent resolution readings, such as 16 at 5X (in the centre) and 8 at 11X (also in the centre). On the other hand, we obtained satisfactory results by producing photographic enlargements from 105mm negatives.

REDUCTION RATIO

Our initial plan was to work with only three reduction ratios: 5X for maps smaller than 19 x 27½ inches; 8x for maps to 31 x 44 inches; and 10X for maps to 40 x 55 inches. It was planned that maps larger than 40 inches and longer than 55 inches would be microfilmed in sections.

After two months of production we have reexamined our plan. Several maps were just too large to be done at 5X. Considering that they would be too reduced at 8X, we have decided to use a 6X reduction ratio. We will probably use others as well. This will not affect our production rates because our maps are filed by size groups. We are accumulating those which are too large and we will microfilm them when there is a sufficient number. This is for the convenience of the operator, who will be spared numerous camera adjustments. All of this is part of our general effort to achieve maximum possible legibility for the researcher.

HEADINGS AND NUMBERING SYSTEMS

Indexing.

We had to establish an indexing system different from our classification system because we do not have a unique call number (e.g. H2/300-1761). Only a small portion of the collection is classified this way. A large portion is arranged by record group concept, and series maps are filed under their own indexing systems.

For these reasons we have decided to assign a unique microfilm number to each map. For example: NMC 00314. This gives us the opportunity of presorting the maps by size and format and filming for several days at the same reduction ratio. However, it does create additional work. We must note this number in five different places: on the back of the map; on the card catalogue; in the register book; and on two closed-storage envelopes.

Institution Identification

Public Archives Canada
National Map Collection
No reproduction without permission

Archives publiques Canada
Collection nationale de cartes et plans
Reproduction interdite sans permission

We are indicating "no reproduction without permission" because we are photographing everything, including photocopies, transcripts, blue-prints, etc. for which we do not always hold the original documents.

MANAGEMENT PROBLEMS

Our programme has suffered several delays caused partly by our lack of expertise and experience and partly by the inevitable "red tape" in any bureaucratic system. For instance, we had to wait several months before the Public Works Department installed drapes around the camera. Second, the reduction of the length of the fiche from 158mm to 148mm was a complex and slow operation. Third, we found out that the resolution readings given by K+E were too low. The camera then had to be recalibrated. Finally, just before starting, we discovered that the lighting was uneven, giving an uneven density on the image.

These problems have been solved but we are still facing others. The processing unit is located seven miles from our office. This means that there are always delays of two or three days before we receive the processed film back. The corresponding maps must be retained until we have checked the density on each fiche. We have ordered a densitometer and a fiche trimmer. This should accelerate the production rate somewhat.

We had to train one of our staff members to be the camera operator. It took two months before he became familiar with the camera and there is still much for him to learn.

Because of staff limitations, the camera operator was the only full-time person on the project. He now has an assistant and we are planning to contract for yet another. With this manpower, we should be able to microfilm 200 or more maps per day.

The first 20 rolls of film developed did not meet the standard for archival quality. All of these rolls will have to be rewashed. And this problem is still not completely solved as yet.

We are still looking for a reader/printer. It seems strange that there is nothing on the market right now when there are so many institutions interested in 105mm microfiche. We are investigating two possibilities. One is to convince Itek Business Products Limited to manufacture a reader/printer similar to the one they marketed several years ago. The second is to ask 3M of Canada to convert the 3M 201 reader/printer to accept 105mm negatives. This second possibility does not seem too promising.

CONCLUSION

There are a few other advantages that are foreseen from our microfilm programme, but which are not necessarily particular to it. First, we are considering microfilming groups of plans that have low historical value. We will then keep only the microfilm negatives and destroy the original documents. In this way, we will save storage space.

The programme should considerably improve our reference services. I have already mentioned our project to build a microfilm room. In addition, large groups of plans now stored in a warehouse at the city limits will become more accessible to researchers. The major improvement will be that numerous maps will not have to be retrieved for a researcher, who, after checking the descriptive cards, thought that such maps were relevant to his topic. Now, just by looking at the microfilm, he will be able to eliminate those maps that are not relevant.

Our programme is still in its infancy. We will continue to improve those areas that are still problematic such as resolution, and the selection of a reader/printer. We will also be experimenting with colour film. I must admit that at one point we had doubts that we were moving in the right direction. However, these doubts and fears are past. The 105mm programme of the National Map Collection is now a reality.

PROGRESS REPORT (November 1978)

A year later, the microfilm programme of the National Map Collection is well underway. More than 20,000 fiche have been produced so far and are currently used by researchers and for photoduplication purposes. We must say that we are still facing technical problems. Production was interrupted twice by major breakdowns of the camera, but we have learned to live with these problems.

In terms of services offered from the 105mm negatives, our plans are realized. Approximately 75 photographic enlargements are prepared each month. The quality is extremely good. And, researchers have available the following enlargement sizes: 15 x 20 inches, 30 x 20 inches, 15 x 40 inches, 30 x 40 inches and 60 x 40 inches.

Prices range from $3 to $18. The maps can be reduced to fit one specific size of paper or reproduced at scale. We have been able to purchase a reconditioned reader/printer that produces good research copies at a low price. This service is available to researchers who urgently need copies. The 105mm negatives are also used to prepare regular prints (8 x 10 inches, 11 x 14 inches, and 16 x 20 inches

No institution has yet started to order duplicate diazo fiche, primarily because the volume of maps on 105mm microfiche was too low to justify buying special readers. However, we do expect to begin receiving orders soon on a selective basis. The high cost of collecting diazo duplicates comprehensively--given the ephmeral nature of many of our maps for smaller libraries--probably means no one else will acquire a complete set. But it is the high cost of viewing and reprographic equipment which will remain the long-run deterrent.

We have checked on the reader/printer to see if good copies can be produced from the diazo negatives. We have compared a copy of the same map generated from the diazo negative with one from the silver negative; the diazo copy was clearly satisfactory. We expect that for most maps diazo will give good research copies, although we are currently using the silver negative for copy production in the National Map Collection.

We now have a year of production behind us, but only four months of use of the microforms in the reference room. For that reason, it is difficult to appraise the reactions of our patrons. So far, their comments are favorable; at least the usual resistance to microforms is not evident. Requests to see the original maps are frequent, but permission is granted only when color is essential or for carto-bibliographical studies. Staff members are also encouraged to consult the microforms instead of the original documents.

In terms of conservation, our goal was to stop the circulation of original maps. We have been able to achieve that major goal. The number of maps sent to our Reprographics Services unit is decreasing rapidly. This trend should continue since we are filming the more heavily used maps first. After completion of the first project, filming small size maps of 20 x 30 inches or less, we did a large group of insurance plans corresponding to our published catalogue, FIRE INSURANCE PLANS IN THE NATIONAL MAP COLLECTION. We decided to put them on 105mm negatives to avoid the accumulation of thousands of photostat negatives which present a major storage problem. We now are planning to miniaturize a group of frequently used oversize maps in the near future.

Now that most of our management and technical problems are solved (or under control), we hope to be able to keep a regular rate of production. The most frustrating thing is probably to see that there are still so many maps in the collection that are not on 105mm negatives as yet and that we are acquiring more maps each year than we can possibly microfilm.

Application of 35mm Aperture Card Technology to Microcartography: How to Acquire a System *

Carl Lundquist, Principal Administrative Assistant, Bureau of Engineering, City Hall, City of Los Angeles, Los Angeles, CA 90012

When a map is stored, two things are placed in storage: a physical object and a complex assembly of information. The physical map is not the information, it is the medium by which the information is transmitted. The information assembly, conversely, is not the map. The physical map itself has intrinsic value, which may range from the going price for scrap, to the going price for a masterpiece. The Declaration of Independence can be found in any seventh grade civics book. The manuscript on exhibit at the National Archives is a national treasure which would command an incredible price, if available in the marketplace.

This diversion into McLuhanesque theories of medium and message is designed to establish that microfilm has nothing to do with the storage or preservation of maps. Microfilm is a medium for the transmission of the information contained in these maps. The application and value of micrographics to map librarianship has to be considered from that standpoint of information transmittal in space and time: in space, to a user denied access to the original by restriction or distance, and, in time, to future users. A microfilm system gains its power as an information system through three properties: scale reduction, replication and stability. Microfilm contains its information in a small, easily handled format, which permits easy duplication and is capable of long life (depending

---

* Presented to the Fall Meeting of the Western Association of Map Libraries, La Jolla, California, October 7, 1977.

on the chemistry of the process used). These properties are not subject to trade-offs with each other. A micrographic format can readily provide great ease of handling and duplication, and possesses archival stability (35mm aperture cards are just such a format). The trade-offs begin when the types of information to be transmitted are balanced off against the properties of microfilm and the pocketbook of the using institution. The more types of information which must be transmitted in a given technology, the harder and more expensive it becomes to transmit it and the more sacrifices must be made to do it.

A map serves to transmit three types of information: digital/grammatical symbols, spatial relationships, and color. These can vary in their importance from document to document. The information types can affect each other and the technological requirements and benefits of a system. Therefore, it is worthwhile to consider them in the order of the demands which each one places on a micrographics system.

Probably the least demanding information for a microfilm system is digital/grammatical data. The limited number of symbols, and the rules of arithmetic and language, permit the loss of considerable data without loss of information. Microfilm systems which handle data that are largely digital/grammatical can utilize relatively low resolution cameras and high reduction ratios. This, in turn, permits the use of either cheap cameras or highly mechanized, high speed ones. Such systems usually have 16mm roll film or 105 x 148mm multiple-image microfiche as an end product.

Information which depends on the reproduction of spatial relationships is far more demanding. Such data may include not only scaled graphics but also detailed renderings and highly organized works of art. It may even include digital material in exotic or highly ornate type fonts, and intricate or unusual hand scripts. For such data, cameras must not only be of high resolution but also of low distortion. The lenses of such cameras are expensive and their mechanical construction precise. Their operation is always manual and slow. Demands are also placed on film quality, accuracy of exposure, and precision of processing. Systems which deal with such data usually employ 35mm microfilm which is formatted into aperture cards. Other applications use roll film on reels or cut into strips and mounted to produce microfiche, while others utilize 105mm film, to the detriment of their information handling abilities.

Color places the maximum demands on a microfilm system. In fact, at the present stage of microfilm technology, no satisfactory color-based system exists. Color-based systems are available but they require major sacrifices in handling, duplication and stability benefits. A color-based system which attempts to approximate the ability of a monochromatic system to transmit spatial relationships is expensive -- 10 to 100 times the cost of a black and white system. Even then, the color film has only half the resolution of monochromatic film. In addition, color film has an archival life dictated by the stability of complex organic dyes. These dyes are sensitive to heat, light, chemical attack and age. Conventional monochromatic film, when properly processed, has an archival life dictated by the stability of purified metallic silver bonded to a polyester plastic base. In fact, it is the only film currently accepted for archival storage applications.

The trade-offs of the above factors against the benefits to be derived in a micrographic system, has, over the past thirty years, made the 35mm aperture card the format of choice in industrial and governmental engineering applications. Mounted in a standard tabulating card, the 1.25 x 1.75 inch (3.18 x 4.45cm) frame permits an image size of 1.97 x 1.615 inches (3.04 x 4.10cm).[1] At a reduction ratio of 30:1, this permits a document 34 x 48 inches (86.4 x 121.9 cm) to be filmed in a single image. Larger documents are handled by sectionalization. The resolution of the best cameras and films available (see below) is very high - on the order of over 160 line pairs per millimeter. Blown back 30X through a precision enlarger, a micro-image resolution of 160 lines/mm gives an enlarged resolution of 5.3 lines/mm: the human eye is just barely capable of that resolving power. My own organization, the Bureau of Engineering of the City of Los Angeles, routinely uses 35mm microfilm to produce "original" drafting masters for plan production. These drafting masters hold scale better than original tracings because of thermal and mechanical deformation of the originals in storage.

Thus, the precision of the 35mm aperture card is fully adequate to handle the digital and spatial information content of maps up to 34 x 48 inches. However, in addition to the precision of the 35mm aperture card, a property which it shares with 105mm sheet film, aperture card users have at their disposal a well established, widespread, mature technology for production and information handling which can take advantage of the technology of tab-card processing. The format is tied to the Federal Government's defense procurement system and is steadily growing in use throughout the public works and heavy construction industry. The technology is so well established and mature that it is even considered a little dull and is rarely the subject of journal articles in the micrographics field.

This technological maturity may be dull but it can also be reassuring to a user venturing into the field of micrographics for the first time. A 35mm aperture card system user has a wide selection of well tested production and use equipment available, both new and used. State-of-the-art production equipment is widely distributed through service bureaus which deal with high technology industries. In short, a potential system user has a wide range of economic strategies available with which to enter the field.[2]

The strategies will revolve around the tasks to be undertaken in an aperture card system. Each of these tasks has an equipment cost. In an absolutely first cabin system, the task and capital cost of each phase would be as follows:

<u>Production Tasks</u>

1. Document filming		$15,000
2. Film processing and inspection		5,000
3. Card mounting		1,000
4. Card labeling		1,500/yr.
5. Card and film duplication		3,500
	First year cost	$26,000

<u>User Tasks</u>

1. Storage		$300
2. Reading		1,000
3. Printing		8,500
	First year cost	9,800

This first-class production system would also require the services of an experienced microfilmer. With one employee, the filming capacity would run about 30,000 documents per year and cost 50¢ per document for labor, equipment (10 year write-off), and materials. These figures may present a clue to the choice of strategies for a prospective user: avoid production unless you contemplate a volume approaching 30,000 documents per year. Fortunately, the alternative, an adequate service bureau, is fairly widespread in the West because of the aerospace industry.

Of the user tasks, only storage and reading are really essential. Storage is the simplest of all problems: the steel tab card file is ideal for aperture card storage. With the gradual abandonment of tab card systems by the data processing industry, used equipment can be obtained easily.[3] For adequate reading, a microfilm reader with an 18 x 24 inch screen and 15x lens is mandatory. Such machines sell for $900-1,000, but, because they are simple, rugged machines with few mysteries, they are good buys on the used equipment market even for an unsophisticated buyer.

Printers are more difficult. They have relatively high maintenance costs and supply costs are critical. The market for these machines is currently undergoing an upheaval with the introduction of electrostatic processes by Xerox and Océ superseding the older, silver based electrolytic and stabilization processes. Fortunately, the use of printers is optional in many systems and can be a source of revenue if prints are sold. Most informational benefits of a micrographic system can be realized with reading capability alone. If a strong demand exists for print capability, that capability can be made to pay its own way easily by print sales in a public service operation. Otherwise, concentrate resources on obtaining an adequate 18 x 24 inch reader.

The importance of good readers cannot be overstressed. Readers can make or break a micrographic system. No matter how good the film is, a poor reader can distort images and reduce their resolution. Poor carrier design, wrong screen size or wrong lens magnification can hamper users so severely that the value of the microfilm is exceeded by the difficulty of using it. The readers used by microfiche systems are categorically poor aperture card and 35mm readers. Their screens are too small and their carriers ill-suited. In fact, this problem is the main limitation on 35mm microfiche and jacket systems: the 35mm image is poorly handled by fiche readers and the 105 x 148mm fiche format is poorly handled by aperture card readers. A potential micrographic systems customer must keep firmly in mind that the objective of the system is not the creation of little chips of film. The objective is, rather, the transmittal of information. It does no good to film if all the technical excellence and effort is wasted by subjecting the ultimate user to a poorly designed or makeshift reading/printing facility. The user will simply fail to utilize the system and the effort will be totally wasted.

The product of a production system, whether done in-house or by a service bureau, will be a film chip mounted in a tab card. The film used should be high contrast, high resolution conventional silver halide with an anti-halation backing, such as the Kodak AHU films. It should have been processed to meet the standard of MIL-M-9868D. The image will be negative (black ground, light lines). At least two aperture cards should be produced per image. One card is for use, the other is for security against loss or damage. A second security card may be produced to double the security factor and permit some use of the first security card. The simplest and cheapest way to produce up to five duplicate cards is to make multiple shots on the initial filming. If a positive image is desired (an advantage with continuous tone material such as photographs and shaded drawings), the image should be reversed by roll-to-roll duplication before mounting and a positive security card retained.

When a card, which has been prepared as above, is viewed, the user will notice something missing -- color. As mentioned earlier, color is one form of information which microfilm transmits poorly. Faced with a collection of colored maps, the potential user of a microfilm system must make a decision. Is the information transmitted by the color essential to the message of any given map? Much use of color is meaningless decoration or redundant coding (e.g. coloring wholly continuous countries). Other use of color may be essential either to the informational content or to the aesthetic content, if this is of appreciable value. If, as in the former instance, the omission of color will little affect the research value, the monochromatic microfilm system described above will be adequate. All digital and spatial information will be transmitted, even if in color, because most modern microfilm is panchromatic.

However, if color is essential to the use of the information, then a problem exists because of the limitations of color micrographic systems discussed previously. Probably the best way to proceed, given the current stage of technology, is the simplest way. Shoot the colored document using a good amateur 35mm single lens reflex camera and a fine grained amateur slide film, such as Kodachrome II. Do the shooting by sunlight or electronic flash. The use of a polarizer on the lens will reduce surface reflections which could be troublesome, especially with flash. Send the film to Kodak to be processed and returned unmounted. Cut the images out and insert the frames into the aperture pockets of Kodak Microseal cards.[4] The result is probably as good a color aperture card of a map as can be produced. Place the cards in the user and security collections, make sure the originals are well protected, and sit back and wait for a major breakthrough in microfilm technology.[5]

The benefits to be derived from the use of micrographic systems have been well established by government and industry. The reference/information requirements of a map can be separated from the physical map. The storage of that map no longer need allow for more than minimal access, and thus, the archivist can concentrate on preservation. Reference to the original document will be reduced to the absolute minimum. Security, circulation and dissemination costs all fall by orders of magnitude. The conflict between access and preservation is dissolved, reducing costs and increasing service.

FOOTNOTES

1. Per U.S. Department of Defense Specification MIL-D-9868D. This is the only existing specification on 35mm microphotography and is the only one which governs the industry.

2. The 105mm unit-image format has a few of these advantages. Marketed by a single company, Keuffel & Esser, the range of equipment for the format is narrow: a single camera, reader, and reader-printer. The use of the format does not seem to be growing. In fact, it may show some signs of obsolescence.

3. Used tab-card files can even be scrounged from a data processing facility.

4. The most common type of aperture card, the Filmsort, uses a pressure-sensitive adhesive rim around the aperture to affix the film chip. The Microseal card, marketed by Kodak, uses two thin plastic sheets to contain the film chip in the aperture. The chip can be inserted by hand. Because amateur film has sprocket holes, unlike most microfilm, the Microseal card is best employed for mounting it.

5. Research at RCA with focussed-image holography has lead, over the past year, to the development of zero-order-diffraction microfilm which does not require laser-beam read out. In fact, it can be read out on conventional unmodified microfilm readers. Resolution is on the order of 300 lines/mm for black-and-white and color and the colors are not subject to fading by heat or age. See M.T. Gale, R.S. Engelbrecht, J. Kane and K. Knop, "Surface-Relief Microimages," The Journal of Micrographics, 11 (3) (January/February 1978), 155-162.

APPENDIX.

INFORMATION RESOURCES

The new micrographics user is faced with a rather bewildering array of options in terms of systems and equipment. Unfortunately, much of the information available to innocents in the field of micrographics comes from sales personnel marketing a manufacturer's line. Another unfortunate source of information seems to be specialists in other fields pushing the particular format or system with which they are familiar. Microfiche systems in particular suffer from this phenomenon. While both sources of information can make valuable contributions, especially sales representatives, their advice should be levened with data from more generalized sources, such as the following.

Cortigan, Daniel M.
1975. Micrographics systems. NMA Reference Series No. 16. National Micrographics Association, Silver Springs, Maryland.

A recent brief overview from a record systems point of view.

Ballou, Hubbard W., editor.
1975. Guide to micrographic equipment. 3 vols. National Micrographics Association, Silver Springs, Maryland. Updated with 1976 Supplement.

The industry bible for equipment. Separated into three volumes: production; user; and computer output equipment. Gives capacities, formats, and prices.

National Micrographics Association.
1967-. The Journal of Micrographics. Published bimonthly, 1967 to present.

A benefit of the $40 per year membership in the NMA, the complete journal from vol. 1, no. 1, through vol. 10 is available on 24X microfiche for $40 to non-members.

National Micrographics Association.
1976. Basic U.S. Government standards and specifications. Revised edition. National Micrographics Association, Silver Springs, Maryland.

Contains the basic 35mm and aperture card specifications under MIL-D-9868D and MIL-C-9877-B.

Nelson, C.E.
1965. Microfilm technology (engineering and related fields). McGraw Hill Book Company, New York.

A dated but comprehensive treatment; still the best treatment of the field of micro-cartography.

Stevens, G.W.W.
1968. Microphotography. John Wiley & Sons, Inc., New York.

A general, technical treatment of the subject of photography at high reductions. Recommended for those with a technical bent who are looking for background on the field.

OTHER RESOURCES

Local chapters of the National Micrographics Association. The Western chapters and their presidents include:

Central Arizona
Judy DeMott
Valley Data Corporation
3001 Indian School Road
Phoenix, Arizona 85017

Columbia
Marge Livermore
Tektronix, Inc.
P.O. Box 500
Del. Station 73-235
Beaverton, Oregon 97077

Golden Gate
David Sanchez
Applied Microfilm Systems
2175 De La Cruz Boulevard
Santa Clara, California 95050

Intermountain
DeLoris Nelson
Salt Lake County Recorder's Office
451 South State
City/County Building #204
Salt Lake City, Utah 84111

Rocky Mountain
James P. Howerton
Microfilm Communications Systems
1645 S. Santa Fe Drive
Denver, Colorado 80201

Sacramento
David Johnson
Zytron Corporation
2020 J. Street
Sacramento, California 95814

San Diego
Wanetta Torbett
Foodmaker Inc.
9330 Balboa Avenue
San Diego, California 92123

SCMA
Gerald C. Mayer
Harris Association Microsystems
1246 S. La Cienega Boulevard
Los Angeles, California 90035

Northwest
Kenneth E. Lee
Kenneth E. Lee Company
P.O. Box 129
Lynwood, Washington 98036

Xidex Institute of Microfilm Technology: a traveling two-day course in microfilm technology put on by the Xidex Corporation. Although a corporate program, the course shows little if any product bias and has a high reputation in the industry. Contact Glen Mezher c/o Xidex, 305 Soquel Way, Sunnyvale, California 94086 for locations. The 1977 price is $195. A course adapted from the Xidex course is put on by the Los Angeles chapter (SCMA) of the NMA. Contact SCMA at the address above. The 1977 price is $100.

A Systematic, User Oriented Plan for Topographic Maps on Microfiche *

Robert Markham, Coordinator, Micrographics Laboratory, Library, University of Northern Colorado, Greeley, CO. 80639

As interpreted by Larry Cruse

INTRODUCTION

The Micrographics Laboratory of the Michner Library, University of Northern Colorado, was established by a grant from the Colorado Commission on Higher Education. Its purpose was to encourage the use of microforms on campuses in Colorado, to provide various kinds of micrographics services, and to prepare microfiche sources. The Lab has also attempted to prepare various special collections, one of which will be discussed here. Finally, our goal is to become a self-sustaining, nonprofit service.

THE PROBLEM

We entered the map microform business as the result of a problem common to many map libraries: invariably, maps were checked out, misfiled, folded, were not returned, or a dozen other excuses. Common to all of them is the fact that at any given moment, the collection is always incomplete. If you have an open map collection, you know what I mean.

One of the libraries of another university asked us if we could microfilm topographic maps for them, specifically the 7.5- and 15-minute topographic

---

*Presented to the Fall Meeting of the Western Association of Map Libraries, La Jolla, California, October 7, 1977.

series of Colorado produced by the U.S. Geological Survey? Could microfilm of some form be made available to provide unavailable maps, that is out-of-print maps, lost maps, and out-of-stock maps? Could microforms of maps help various kinds of users, such as professional geographers, novice students, hikers, travelers, canoers, businesses seeking material on censuses or boundary data on counties?

Could microforms provide better access? Most libraries file maps alphabetically. You probably all know that there are two alphabetic systems of filing: the USGS files maps letter-by-letter irrespective of word breaks; libraries normally file maps, like everything else, word-by-word. There is a great difference in some cases between the two filing orders, viz. the filing order of San Ysidro and Santa Barbara: the USGS reads SANT, SANY; libraries read SAN, SANTA. In addition to that, most students know only a place name. This is not necessarily a quadrangle name. If the coordinates of the place are not known either, how do you find the proper quadrangle?

Finally, could microforms help reduce the costs of storing, retrieving, and keeping maps up to date?

THE SOLUTION

Perhaps there are others, but I know of only four means of microfilming maps: roll film; aperture cards; microfilm jackets; and microfiche.

Roll film is the type used by the National Archives to film the USGS historic file and annual updates. They have now filmed a complete set of all maps ever published in these series: 1:250,000 (1 x 2-degree); 1 degree (1:250,000); 30-minute (1:125,000); 15-minute (1:62,500), and 7.5-minute (1:24,000). They are available on 200 rolls of 500 frame microfilm for about $20 per roll. It is an extremely good set of black and white images. For various reasons, we did not choose this system.

Aperture cards: although they have many values, this is the second option we did not choose.

The microfilm jacket system was rejected for one very specific reason: in our tests, we found that the use of jackets reduces the duplicating quality of the film by 30%. We gave a jacket company our film, which resolved 149 lines per millimeter (L/mm) in the master copy. The duplicate made through a jacket resolved 105 L/mm. This is not bad quality, but we rejected the system for that reason, primarily.

We chose microfiche as allowing a flexible arrangement of format and providing the best eye-legible headings. They are a standard size, easy to use in libraries, easily duplicated on diazo microfiche duplicators, and are easy to find and refile.

THE RESULT

The result was, we produced a complete set of customized microfiche. Each microfiche contained eight maps and had an easily read eye-legible heading which contained a rather unique arrangement of information. The set also contained a comprehensive map-name index. The maps were filmed in cine mode, with the long side of the map parallel to the long side of the fiche. This was the only orientation possible at the time that would allow a 20X reduction.

When this first set was produced, people immediately complained. Although maps are filed alphabetically in most map collections, it is done solely out of convenience: no one uses them that way. Consequently, we changed the map arrangement.

Our second attempt with the 7.5- and 15-minute quadrangles formed a geographical sequence. The eight quadrangles on each fiche are eight adjacent quadrangles. This gave us some new possibilities.

Each fiche consists of two 35mm film strips which form a perfect 105mm chip of standard microfiche width. It is our own unique, unpatented invention. To create it, we took two parallel strips of double-sided masking tape laid edge to edge to form one adhesive strip. The film strips were laid on these and then a special kind of scotch tape was placed over the top, creating a duplicating master. A third parallel strip of 35mm film was joined at the top by the same means and carried all of the header information. This master never leaves the Lab. The duplicates from it, which are all anyone ever sees, are generated by the diazo process which leaves the edges of the tape invisible. The scotch tape is very strong, will not tear, is relatively easy to work with, and produces a fixed, standard size microfiche.

The reduction ratio of this set is 21X. The motivation for the original cine mode was its allowance of a 20X reduction. Changing formats forced us to reduce the image slightly to keep it within the 4 x 6 inch format, which is about 1/2 inch too small otherwise. A 20X reduction is almost possible, but is very much more difficult to achieve. With that reduction, the maps will almost be touching, which has both strengths and weaknesses.

The original eye-legible header strip had some defects also. The top line provided the source, "USGS TOPOGRAPHIC MAPS," the name of the state, the date it was filmed, and the microfiche sequence number. There were 194 microfiche containing maps of Colorado. The second line gave the quadrangle name and date.

In the geographic set, the title strip is negative, as are the maps, so that both have white lettering on black backgrounds. Duplicating it on a reader/printer reverses its polarity; the result is an all positive product. The information it contains includes: series name; state; filming date; reduction ratio

(21X); latitude and longitude; range and township; Army Map Service (AMS) number; map name, and map date. The AMS number is a great aid for geographical finding and filing purposes since each fiche contains two AMS number sets (in the 15-minute series) or subsets (in the 7.5-minute series). Since our first goal was to make the maps more usable, all of this header information furthers that end by giving us multiple addressability. Few, if any, manual filing systems allow for so many approaches.

The result, now, is a complete set of 1,500 topographic maps on 194 microfiche. In our library, these same maps on paper occupy 4 file cabinets, a total of 20 map drawers. They are hard to get at because of their weight. They are accessible only by a topographic index while our comprehensive microfiche index gives us access by every variable on the header strip.

DIFFICULTIES

Creating this microfiche set entailed certain difficulties. Since our collection is an open one, it was impossible to find a complete, sequential set of maps to film at any given time. Fortunately, we live near the USGS's Western Distribution Center in Denver, which has a supurb collection. Also, Denver has about six map stores, all of which have become very familiar to me. Even so, there were always some maps we could not find.

We had our worst difficulty preparing title strips. Even our present method, with the advantage of experience, is inadequate. Although they are legible, they would benefit from qualitative improvement. We use a varityper on a "Molex Strip," which produces a carbon image, adhesive header for each microfiche master.

Another difficulty was film size. We were limited to our available equipment, an NRD-2 35mm planimetric, microfilm camera. The paper maps we were trying to film were often curled from use. Holding them flat was very difficult with this camera.

We had tremendous problems with image density, but have finally solved them. These problems sprang from the wide range of topographic terrain types on the maps, from almost contourless barrens to adjacent, heavily wooded alpine areas.

SYSTEM STRENGTHS

The strengths of the current microfiche map set are many and varied. Most appreciated by a map librarian might be the fact that they store in such a small space and, unlike paper maps, they file standing up, so every map in the set is equally easy to file and retrieve.

I think they are weatherproof. I am told that diazo fiche is less affected by moisture or humidity than silver film. Paper copies, made on silver base paper in a reader/printer, are very prone to fading. Sunlight turns the entire print chocolate brown. The simple expedient of Xeroxing the silver base print, however, will eliminate this propensity since Xerox's are electrostatic and not light sensitive.

The microfiche set allows for other students, or a number of students, to have access to the same maps simultaneously. Other libraries can buy the maps also.

Maintenance costs is always a very elastic topic, depending on how one wants to measure floor space, staff time, replacement costs, etc. First off, our maps can be retrieved more quickly: alphabetically, by map name, or geographically by place. Since all of the fiche in the set are numbered sequentially, they are easy to find and refile.

The entire set is portable and manageable. It is also very complete.

A limitation of maps on roll film is that only one person can use the roll at any given time. This is inadequate for use in an active collection, the likelihood of simultaneous need rising directly with the number of sheets on the microform. For the USGS project, maps will be in use 500 at a time. With the microfiche, they are in use eight at a time.

Another limitation of roll film maps is addressability. For manual scan, finding any given map can be time consuming, the finding being done after the roll is on the reader. With the microfiche set, the map is found before it is placed on the reader, reducing search time.

The microfiche maps are inexpensive and freely available, in units of 8, not 500. You more nearly buy only what you want, rather than a quarter of the state.

Convenience is another strength. All of the maps are at one place at one time in one box or drawer. They are filed numerically, which holds down filing errors.

Using the two microfiche map sets together, the alphabetical and the geographical, one could have access by map name, alphabetically, by latitude, by longitude, by AMS number, or by range and township number in most of Colorado. Some places in Colorado simply have no range and township number.

Storage of the maps is easy, simple, quick, and compact. This compactness, along with their economy, convenience and accessability may alone point to microforms as a possible alternative for map storage in libraries.

Colorado adds about 250 7.5-minute topographic quadrangles per year. It is very simple, in the geographic set, to add these because they are usually adjacent maps. Where they are not, we only have to add the new quadrangle to the affected subset fiche. In the alphabetic set it is a different proposition. Every time a new quadrangle is issued, the alphabetical set is obsolete and must be completely redone, or a supplement issued, which defeats the objective of having them alphabetically arranged. Supplements actually make it a numeric (year) system with an alphabetical sub-system. This is the case with the USGS microfilm roll program.

Sometimes, too, the map names will change. This is of no real consequence to the geographic set since the coordinates remain constant. Once again, the sets on rolls become obsolete, as does our alphabetical microfiche set.

A third area is map updates. Here the problem is the same for both sets on microfiche. This is one reason, apparently, that the USGS is using rolls of film issued on an annual basis. But it is a very undependable approach: the edition date and the date of actual release are often at variance, as anyone receiving depository shipments knows.

Resolution on the microfiche sets is very good, about 149 resolvable line pairs per millimeter on the master film. At full blow back, this yields a resolution on the order of 7 lines/mm. Human visual acuity, at normal reading distance, is on the order of 10 lines/mm.

<u>WEAKNESSES</u>

There are also disadvantages, let me not minimize that at all. Librarians know that microfiche is not popular. I am a rabid fan of microfilm, but the students are not.

It is black and white. It takes a very careful person to use these maps instead of colored ones. The uninitiated cannot distinguish between contour lines and rivers. If you are looking for rivers, you can find them, they are simply black and white lines.

They are also negative, which is easier to read, but different from the positive paper originals. The reader/printer reverses this polarity so that they print in the positive mode. But negative maps are confusing to most users, at least at first.

The first set of maps were filed alphabetically, which means tomorrow they are out of date. When a new map is issued, the sequence is immediately wrong. It is impossible to expand internally.

The maps are different dates. Although this problem derives from the paper map originals, it is more apparent when the maps are on microfiche in blocks of adjacent quadrangles. Trying to follow a highway from one sheet to another and having it suddenly end is very disconcerting.

Viewers rarely provide full-frame, full-size blow back. The larger the discrepancy between the reduction ratio of the image and the magnification ratio of the viewer, the less well the image can be seen. I have been unable to find a full-frame library type viewer, one in the $200-$500 range. Thus, you will be limited to partial map viewing at full blow back on equipment most libraries already have. Architectural/engineering reader/printers, such as the 3M "200," are available, but prices for such equipment start at $3,000 and go up from there. They do have the advantage of projecting an entire 35mm image on their screens. Microfiche inspection units, in the same price range, while they do not print, can view an entire microfiche at one time and have a second lens capability, which would allow them to view a full 35mm frame, or a portion of a full microfiche image in greater detail.

Finally, we tried color microfiche. In my opinion, it does not yet work well. While it is beautiful to look at and adequate to use, the resolution is less than 50 lines/mm. To the novice, that may mean nothing, but if you are interested in map accuracy and quality, it means a great deal. If the reduction is 20X, the blow back resolution will be on the order of 2.5 lines/mm. I have made a few color samples but the problems were too multitudinous compared to black and white film.

Density on the two films is not comparable: it is almost an irrelevant term with color. A 20 volt difference in a color step test may yield very little change. This leads me to believe that color film is grossly less sensetive.

You can expect about a 70% loss of resolution. But, compensating for that, you retain much more, namely color. For some persons and uses, color is absolutely essential, such as in finding the light blue stream beds on topographic maps.

Normally, the color film is 5 to 10 times the cost of black and white film. I tried for four months to obtain such a roll -- just one roll. Kodak would sell me 20 rolls for $1,000, but not one roll. I think I have solved that problem now, but it took a letter to the president of Kodak to get the film in my hands, finally. I do not know why, but it did.

My second bad experience was trying to get the film processed after I shot it. It was first sent to Dallas to be processed. They sent it to Rochester and then it came back to me, undeveloped. Four months later, they found a place that would process the film. The film came back from this adventure totally blank. I began again at square one.

Now, there are many faults involved, some mine, some others', but this entire comedy is exemplary of the whole new range of problems endemic to color microfilm. I think the future direction must be towards color, but, at this point, the route has many pitfalls.

THE INDEX

The index we found is one-of-a-kind. The staff in the sales office of the Western Distribution Center of the USGS in Denver prepared an exhaustive list of map names on Colorado topographic maps, including the maps we filmed. We borrowed the index from them, put it on microfiche, and now add a copy of it to all map sets we sell. It includes 10,000+ names. Many of these do not occur in any gazetteer, Zip Code directory, or telephone book. They are simply topograpic feature names which are on the maps and are not listed anywhere else. Included are 3,000 mountains of various elevations, 250 mountain passes, and, when it rains, roughly 2,500 named rivers and streams, as well as numerous lakes and reservoirs.

The Map Sales Office staff did this on their own initiative, between selling things, over a period of years. It seems that they took every map, went through the names and made a punch card for each. It is a first rate, beautiful index. When existance of the index became known to their supervisors, the staff was reprimanded.

OTHER PROJECTS

The first topographic set spawned other projects. We completed a set of the 14,000 foot peak quadrangles of Colorado. These 53 maps are on 7 microfiche. They were done for the anniversary of Colorado's statehood and for the U.S. Bicentennial.

Next, we filmed all of the "special" maps of Colorado done by the USGS. These included some Front Range maps, some old mining maps, and a number of maps of erratic size or erratic nature. Now they are all available in one place in our sequence of microfiche.

Then we assembled all of the National Park, National Forest, National Monument, and National Grassland maps and filmed them in the same format as the previous series. All of these maps are contained on 12 microfiche. The set brings together the disparate mapping products of the USGS and USFS (U.S. Forest Service) in one place.

We recently completed filming the complete 1:250,000 scale National Topographic Series, including the subset covering Alaska. This project raised entirely new problems because of the great detail involved. To my mind, the result is less satisfactory than the 7.5- and 15-minute quadrangle sets.

These maps are filmed four to a fiche. The headers show which IMW (International Map of the World) quadrangle they are in, and their sheet number, i.e. NJ-18-6. We also show latitude, longitude, filming date, reduction ratio, sheet names, and sheet dates. Perhaps this is too much information for some, perhaps not enough for others. Since it is eye-legible, it tells you what you want to know before it is placed in the reader. With roll film, this would be prohibitive.

We split the long, coastal strip maps in the series into top and bottom halves, leaving some overlap. They are now properly positioned and properly orientated with respect to the adjacent maps. They are not cut, so you see a complete map in the position it falls on the land, geographically.

When the series was filmed, we were confronted with the fact that there is no index to these maps, outside of the Colorado index mentioned earlier and one for Wyoming, which is the best in the country by far, at least for these maps. Everything on all of the Wyoming sheets is indexed: geysers; tunnels; bridges; airfields -- everything.

We are currently working on our own index to accompany the National Series on microfiche. It will list all National Parks, Forests, Monuments, Indian Reservations, and a few more things, as well as giving the sheet name, its date, scale, and geographic coordinates. It is a print-out at this point but will soon accompany sets of the National Series of 600 maps on microfiche, all in one handly little compact storage unit.

The index was created by first listing all of the sheet names of the

National Series. Then the maps are filed in state order. Once this is done, the data cards return in state order. Third, all of the National Forests, Parks, and Monuments are listed. Again, the same basic information is entered on the tab card, it is run through again in a different order and the list is gradually built up. Finally, we will generate a list of all of the maps in the system. For example, Badlands National Park will show three different types of maps: 1:250,000 series sheet(s); a Badlands National Park map, 1:60,000; and all of the relevant quadrangle maps.

In one sense, it will be an index to the National Parks, Forests, Monuments, and Indian Reservations that is available nowhere else. We are deliberately excluding military installations on this index, at least for now.

COSTS

Finally, let's talk about costs. A complete set of Colorado topographic maps, if available, would cost about $2,000 in paper copy. Our microfiche set of the same maps is available for $90. The index is free. The new diazo film costs us 4¢ per fiche in the raw state; we charge 50¢ per fiche. The difference is used to cover our overhead, about 25¢ of which is directly attributable to generating the microfiche, and 21¢ for indirect costs.

CONCLUSIONS

In the process of this work, I have discovered the three basic laws of map micrographics.

First, people prefer paper. Always, invariably, without exception, people prefer paper.

Second, computers are quicker. They are more "glam rous." Everything is better on computers.

Finally, colors are, I believe, almost essential for map work. I say, almost. I think our maps have many values, many good qualities, but color is not one of them. At times they can be used to accomplish many things that cannot be accomplished by any other means on maps. They are always at your fingertips which, but for its lack of color, has many, many merits.

The Standards Committee on Cartographic Microforms - a Progress Report *

Stanley D. Stevens, Map Librarian, University Library, University of California, Santa Cruz, CA 95060

INTRODUCTION

In mid-1974, the National Microfilm Association (now called the National Micrographics Association, NMA) established a "Committee on Maps" under the general chairmanship of Ralph E. Ehrenberg of the U.S. National Archives.

The objective of the committee, now named the NMA Standards Committee on Cartographic Microforms, was and remains today the development of an industry standard that would become an international standard for the microfilming of maps.

The route of travel between the Committee and the International Organization for Standardization (ISO) is a long and methodical one, requiring adoption on several levels. Following completion of the Committee report, it will be forwarded to the Chairman of the NMA Standards Board who will review it and either approve it for publication for industry comment or make recommendations for changes. If approved by the Standards Board, copies of the proposed standard are distributed with a letter-ballot to the entire NMA membership for comment and vote. NMA, with a membership of more than 8,000 individuals and 250 trade member companies, has 18 committees directly involved in developing and writing standards in such areas as computer output microfilm (COM), terminology, drafting, microfacsimile, equipment, materials and supplies, and

---

* Presented to the Fall Meeting of the Western Association of Map Libraries, La Jolla, California, October 6, 1977.

microfiche. Committee members represent both users and vendors and are persons with substantial interest and technical expertise in the various fields appropriate to the standard committee on which they serve. Through these committees NMA works toward achieving voluntary acceptance of the standards in all segments of the micrographic industry before they become national or international guides.

If a consensus of the NMA Membership is reached, the Committee report is published as an NMA industry standard. Following this, the industry standard is transmitted to the American National Standards Institute (ANSI) and the ISO for consideration as National and International standards.

It should be noted at this point that these standards are based on consensus and are not enforceable. They are recommended for use by the industry, and are acknowledged as practices that should be followed. Usually, however, government agencies require as part of the contract specifications for goods and services that these standards be followed.

THE COMMITTEE

The committee initially consisted of the General Chairman, Ralph Ehrenberg, John Wolter of the Geography and Map Division of the Library of Congress, Stanley Stevens representing the Special Libraries Association, Betty Kidd, Chief of the National Map Collection, Public Archives of Canada, David Woodward of The Newberry Library, John Simmons of the U.S. Geological Survey, Raymond MacNamara of the U.S. Defense Mapping Agency, and James Knight representing the Association of Canadian Map Libraries.

Ralph Ehrenberg subsequently resigned from the Committee and was replaced by his deputy at the Cartographic Archives Division of the National Archives, Charles E. Taylor. Joining the Committee were David Carrington, Head of the

Processing Section of the Geography and Map Division, Library of Congress; Robert Brandt, Head of Research and Development, Advanced Technology Division, Defense Mapping Agency; Joseph C. DiGiovanni, Synectics Corporation of Rome, New York; Dr. Ernest P. Taubes of Potomac, Maryland, and Gilles Langelier of the Public Archives of Canada (who replaced James Knight).

Meetings of the Standards Committee have been held in Washington, D.C., New Orleans and Chicago. Three preliminary drafts of the standard have been produced.

THE STANDARDS CONTEXT

The proposed standard has been limited to filming procedures for microfiche, 35mm, and 105mm microforms. Seventy-millimeter filming procedures have been excluded from consideration since ISO has dropped 70mm as an international standard - too few people are using it at the commercial level. Film processing, film storage, and film retrieval have been excluded because adequate standards for these procedures already exist.

Subcommittees were formed to develop systems-standards for each film size: the 16mm subcommittee was chaired by David Woodward, the 35mm subcommittee by Stanley Stevens, and the 105mm subcommittee by Charles Taylor. Elements considered for the system-standards were: index information (or heading), film format (roll, film chip, etc.), image sequence, reduction ratios, film mounts, and recording medium. It was decided that one set of quality and inspection standards would be prepared for all three film sizes, but each subcommittee would consider resolution factors, density of image, archival quality, duplicating ability, and filming procedures. The orientation of frames was considered, along with sectional filming of documents too large for one exposure, frame identification, and bibliographic information.

The following features were considered necessary for the proper achievement of consistently satisfactory images: camera quality that can meet or exceed minimum legibility requirements, inclusion of a reduction bar within each image to facilitate identification of the specific reduction used (for blowback viewing or printing), and the inclusion of a gray scale, color scale and resolution chart placed in the image area.

The standard will deal with both continuous-tone black-and-white silver-gelatin type film on a polyester base, and color film (although I do not expect many projects to be filmed using current color technology).

OTHER STANDARDS

Many of the specifications are merely references to other standards which have already been adopted, so that the essence of this new standard (when finally approved) will relate primarily to methods of filming cartographic materials.

The physical characteristics, size, type and format of the film, for example, have already been established. There is no need, or desire, to work outside of those standards.

TESTING THE THEORY

Extensive tests and experiments have been conducted simultaneously with the development of these standards, so that it will be known whether theory and practical application conform before the standard is adopted. Charles Taylor has conducted a Map Microfilming Feasibility Study at the National Archives, and other members of the committee have demonstrated models of their recommendations.

CURRENT STATUS OF THE COMMITTEE WORK

At this point, the Committee's recommendations are entering the final stage of drafting. Review of the final draft is underway, and it is hoped that map librarians, archivists, microfilming companies, and users will soon have a standard on which they may rely.

Although one of the first materials ever to be microfilmed, maps are one of the last formats to receive their own filming technique standards. There have been many filming projects in the past, particularly by military agencies for their own use, but with the adoption and publication of formalized standards, we can expect to see map microfilm projects appearing regularly on the market. These will include both historical materials and contemporary files.

Can't you imagine the entire set of USGS 7 1/2-minute quadrangles for the entire United States -- filed in a single 5 x 7 inch shoe box?

POSTSCRIPT (August 1980)

The National Micrographics Association has reorganized several of its standards committees. The former NMA Standards Committee on Cartographic Microforms, which was chaired by Charles Taylor, is now a subcommittee of NMA Standards Committee C-7, (microfilming of engineering drawings, chaired by Lester O. Kruger). Ralph E. Ehrenberg, formerly of the U.S. National Archives and now Assistant Chief, Geography and Map Division, The Library of Congress, has resumed the Chairmanship of this subcommittee.

New to the membership of the committee are: Larry Cruse, University of California, San Diego; Albert J. St. George, California Computer Products, Inc., Anaheim, CA.; James Becker, Photographic Sciences Corporation, Alexandria, VA.; and, Tony Salerno, U.S. Geological Survey, Reston, VA.

The new subcommittee will be known as the NMA Cartographic Subcommittee, C-7.3, of the NMA Standards Committee, C-7, Microrecording of Engineering Graphics.

Commercial Reprographics Services*

Bill Talbot, Sales Representative, San Diego Blue Print Company,
4696 Ruffner Street, San Diego, CA 92111

As interpreted by Larry Cruse

San Diego Blue Print Company, representative of reprographics service companies in large metropolitan areas, is in the business of taking hard copy documents and reproducing them to client's specifications microphotographically. The same services are available, of course, if you obtain cartographic microforms elsewhere and bring them to us for such processing as enhancement, enlargement, reduction, blow backs, and printing on paper or plastic.

The 105mm microfilm system we work with is manufactured by Keuffel & Esser Company (K+E). It is unique enough that many people find uses for it, especially engineers, architects and similarly engaged professionals putting cut-and-paste work together using bits and pieces of old drawings, creating new material without resorting to pencils. Basically, this is the kind of work the Micromaster Camera was designed for. But it can also multiply the utility of maps by microfilming them and manipulating the resultant micro-product, just as is done with architectural and engineering drawings.

We prepared several 105mm film negatives of various maps loaned to us for that purpose. In most cases, the work was quite straightforward, in others it was somewhat more difficult. Using these as microform masters, we then generated

*Presented to the Fall Meeting of the Western Association of Map Libraries, La Jolla, California, October 7, 1977.

a number of products to demonstrate the flexibility and fidelity possible with well prepared cartographic microforms.

Generally, when we photograph the original map with the 105mm camera, we place a scale on the copy board or camera bed. This scale appears across the bottom of the resulting negative, becoming an integral element of the microform. In this way we can always tell the reproduction's size relative to the original, and we can achieve requested blow-back ratios with little difficulty.

We can also photograph header information simultaneously with the map, although you are probably better off so far as eye legability is concerned to apply the header information yourself, after the fact, with an adhesive titling strip produced on an office typewriter. If you so specify, we will allow for this space to remain blank, keeping the map image from encroching on that area of the negative.

You now possess the basic ingredient of a microform map collection. You also have a readily exploited base map whose use is limited only by currentcopy-right restrictions. Although it is in black-and-white, you have the reduced equivalent of a copy negative of the original map and are free to produce a number of nonconflicting, positive print alternative products from it, depending on your or your patron's needs. All of these alternatives--especially those impossible on office equipment--can be realized through a service agency. This makes it practical (if not convenient) for you to rely on a microfiche reader as your lone piece of departmental equipment, at least at the start.

One product your patrons might be interested in is a standard 8 1/2 x 11 inch or 17 x 22 inch positive print on paper for inclusion in reports, theses, monographs, or as a master for making multiple Xerox copies for class use. Note tha with the addition of a little "pen and ink" on the first Xerox print, a completely new map is created without compromising the microform or the master print. If the intent is to make ditto copies, the first Xerox print is useful for that, too.

There is really no limit to the permutations possible when utilizing a little creative Xerography and a positive print from a negative microform.

The same microform negative could also be used to generate a print on stable based plastic for use as an overhead projection transparency. This can be multi-overlaid with annotations, other maps, etc., just as is done in the sets sold by educational supply houses, except here everything is done exactly as your patron prefers. It's similar to the difference between buying clothes off the rack and having them custom tailored. The transparency can serve double duty, too: it will Xerox just as well as a print on paper, with the additional advantage of being gangable, either by stacking and Xeroxing multiple overlays together or by running the Xerox print through the copier for multiple exposures, once for each overlay. In the latter instance, registration can be difficult; in the former, there is a practical limit to the number of transparencies the Xerox can keep in focus. The whole process can be reversed almost as easily by using a white opaque liquid on the Xerox print to mask out unwanted information before reXeroxing. And do not forget, all of the same Xerox manipulations can be done on a color Xerox so that the base map is, say, one color and each overlay some other. Felt tip markers might be interesting to experiment with, just to have a sample you can show to patrons as a guide to the various things they can produce with your cartographic microforms, our prints and a Xerox machine.

A third product which I should mention is a full size print of the original map from the microform negative, printed positive, to be used in conjunction or in lieu of the original map for whatever purpose your patron has in mind. Such a black and white reproduction need not be thought of as something less than the original colored map necessarily. In many cases, color on a map being used for secondary purposes can be considered "static" which confuses the user. Often, it is better to use marking pens of different colors to emphasize those elements you want noticed on the reproduction.

Reproducing the original map from the microform master onto a polyester base material allows additional latitude over paper prints. Here, two maps can be directly overlaid since one is transparent. If their scales were different in the original state, the polyester overlay print can be reduced or enlarged to eliminate the problem. Unfortunately, the transformation of projections is beyond the camera's ability, but once the scales are reconciled, differences due to projection differences are often inconsequential, especially with large-scale maps. But this is the kind of problem which is easily assessed by projecting the microform onto the map it would overlay to test its "fit" before ever committing yourself to such a print. Such overlay techniques are especially useful in doing cultural or environmental change studies.

Another example of of this type of treatment, one with which you may be familiar is the Bodleian Library's nomenclature overlay for their color facsimilie of the Gough Map of early Briton. Another variant is the U.S. Geological Survey's land use overlay series done in conjunction with their 1:250,000-scale topographic maps. There is also the possibility of using the overlay technique as a planning medium, anticipating the "fit" and "look" of a proposed change to a natural setting. through selective deletions and inserts.

One illustrative difficulty we overcame with the maps we microfilmed had to do with the quality of color involved, not color itself. On some of its maps, the U.S. Forest Service uses a particularly dark green over large portions of the mapped surface. Along with the dense oranges and reds they use, these greens photograph black. Maps originally designed with microreproduction in mind would have all of these colors done in pastels, tones which can easily be controlled. But a very dense heavy solid area, such as the Forest Service's dark green, forces us to use a mechanical screening process which, when placed over the original map, serves to "soften" the dark areas by breaking them up into a dot pattern of preselected density. Unfortunately, the areas which were not a problem before certainly are now. Pure white areas on the original are now a bit grey and look somewhat inferior. This problem can be overcome of course: simply make a screened and an unscreened print, cut and paste a new map from them, then microfilm that.

There is always the possibility that one of your patrons would like to take a color microform map, or its paper equivalent, and make a color facsimile of it. Service agencies such as ours with four-color process printing capabilities can do that also as we have with a rare map given to us by the Cabrillo National Monument staff. Qualitatively, I am not sure the USGS is doing any better work.

I assume that you, as librarians, are interested in how you might financially exploit the economies of map microforms to your library's advantage. There are two approaches. First, you can order your own 105mm "Micromaster Camera" from Keuffel and Esser for $18,000. That is the price of the camera alone and does not include the necessary support equipment such as processing tanks, nor the desirable options which are available, such as a 35mm conversion kit. Or, for $1.90, San Diego Blue Print will do the negative for you. This price holds for each of the first nine black and white negatives. For ten or more, the price drops to $1.70 each, and for large volume the unit price can drop as low as $1.30 each. Included in the basic price is an archival envelope or sleeve and normal opaquing or touch-up work.

So far as 105mm color microfilm is concerned, I must admit that San Diego has not become sophisticated enough to create a market for it, so we have not established a price schedule. We know we can do it, however, because we have run color tests.

Paper blow backs are $1.05 per square foot. The polyester base drafting material, which is translucent but not transparent, or the clear film is $2.90 per square foot. While that may sound like a lot of money, it is considerably cheaper for a custom product of your patron's choosing than is $18,000 for a camera plus the cost of a technician to run it, expending the time it takes to trace a map by hand, or to draw a completely new one.

I have not mentioned 35mm microfilming because, in the microreproreproduction of large maps, 35mm imagery is simply not adequate for regenerating the material back to full size in a clear and accurate way. We do have 35mm capability and we use it a great deal, but for this type of work I would not even use it experimentally. You can do a great many things with 35mm, but you would be better off limiting it to engineering drawings rather than subjecting it to the demands imposed by large, complex maps.

The Manuscript Map Microfilming Project of The National Archives *

Charles Taylor, Assistant Director, Cartographic Archives Division, National Archives and Records Service, General Services Administration, Washington, D.C.

As interpreted by Larry Cruse

The Cartographic Archives Division of the National Archives of the United States has been experimenting for several years with the microfilming of manuscript maps. Usually, these are unpublished pen and ink drawings anywhere from one hundred to two hundred years old.

In a feasibility study several years ago it was concluded that the only way we could film these priceless maps was to use the 105mm format, because of the condition of the records, the paper involved, the line widths and weights, and the lettering sizes. Of course, we could have utilized an 8 x 10 inch copy camera if we had wanted copy negatives, but this size would have been unnecessarily large for our purposes and we would have lost the advantages inherent in the microfiche format.

Thirty five millimeter was judged inadequate because it was unable to resolve all of the up-to-40 different lettering sizes appearing on some of the maps. These sizes ranged down to the level of two point type, about one-half the size of phonebook type. The letter "e" on such manuscript maps, even when using the two point, was not filled in. When utilizing 35mm microfilm, these "e's" were invariably filled in, unless we over-sectionalized the record.

---

* Presented to the Fall Meeting of the Western Association of Map Libraries, La Jolla, California, October 7. 1977.

Some of these records measure about 40 x 50 inches. If we had sectionalized them at 35mm, we might have needed as many as eight sections to get a satisfactory result. Using 105mm, we were able to capture such maps in one image, although we were near the limit of the system's capabilities.

About one and a half years ago we decided to do a pilot study. This would allow us to look in depth at the capabilities of 105 mm microphotography for black and white and color reproduction. The result was an approximate total of 9,200 individual film pieces (two each of color and black and white for records photographed). We felt that was enough to give us an idea of the strength of the system. We especially wanted to explore the characteristics of color film for reproduction of maps, feeling that the researcher who came to the archives would be interested in it if the map were color-coded or color-keyed.

A group of maps was selected which contained examples of all of the types of map microfilming problems which were anticipated in the overall manuscript collection. The black and white film selected was Kodak type 2460, a polyester film with high resolution capabilities and good dimensional stability. The color film was Kodak SO-356. This is a direct, positive Ektachrome film, originally an aerial photography film. It had the highest resolving capabilities of color film then available and I believe that this is still true.

At the start of the pilot project we used the engineering specifications developed by the U.S. Army. This resulted in a film chip size of 101.60 x 152.4 mm. Our current film chip size is 105 x 148mm, down from the military specification. We then allow for the header area which reduces the effective image-carrying surface to 140 x 105 mm. The camera lens was a 210mm Nikon 5.6 enlarging lens. The three reduction ratios were 5.5 and 8 and 10X. When we changed the information carrying area of the microfiche we were forced to change the reduction ratios accordingly. The reduction ratios are now 5.3, 8.5, and 10.6X. These multiple reductions were necessary because our filing system at the National Archives is one of provenance, not size, as is the case in some other archival collections. We also wanted to film the maps in sequence because we did not have the manpower to rearrange them by size (the three reductions allowed us to do that).

We used a superentitling camera to provide the heading information rather than use a band placed on the copy board. This camera provided title information across the top of each chip. In the titling area we used the National Archives coding system: Record groups, agency name, the agency's name for the file containing the map, and if applicable, the map's number.

The copy table used was 42 x 72 inches. It incorporated a large sheet of polyester material to hold the maps down, in conjunction with a vacuum system. This is fairly standard practice when dealing with oversize material.

Other specifications for the filming included precise instructions for the orientation of the maps on the copy board. We had to sectionalize some records and if necessary we would turn the maps ninety degrees to fill the available area of the copy board. Sometimes this was occasioned because the maps were basically north-south oriented, rather than east-west. We also specified overlaps when sectionalization was necessary.

The resolution requirements for both black and white and color film at full blow/back were 12.5 lines/mm in the center of the National Bureau of Standards Resolution Chart and 11 lines/mm in the center and 9 lines/mm in the corners. For 10X reduction it was 8 lines/mm in the center and 7.1 lines/mm in the corners. These were the minimum allowable readings.

The test pattern was photographed at the beginning of the day's filming. It was also shot at the midday break and then again at the end of the day. Consequently, these test patterns were available at 3 different stages of the day's production. The target was read with a 50 power microscope and this gave us the particular values, using a pattern definition method which is standard procedure in the field.

I think the method of testing the resolution was satisfactory, but once in a while we had a deficiency in the resolution because of some particular problem in getting the maps as flat as necessary on the copy board. Part of the problem we attributed to the fact that we were dealing with deteriorated maps which, while not preserved folded in the Archives, had been folded at one time. We can take some of these folds and wrinkles out of the maps, but we had to do it very carefully. Even so, we often had problems getting the maps flat. If we could not get them flat on the copy board we were not going to succeed in obtaining high resolution over the entire face of the chip. Consequently, the record had to be rephotographed once we knew where the deficiency was.

We also wanted to use a special test target, the RIT (Rochester Institute of Technology) Alphanumeric pattern; it allowed us to test the accuracy of our reduction ratio since we could measure the letters. This is something we could not do with NBS target since it is based on the separations between lines.

The density requirements were then determined by the density of a 50% reflectance target. We also used 95% reflectance targets. The density values for black and white and color film at 95% reflectance was a range of .90 to 1.3 and for 50% reflectance it was a value range of .10 to .3. One of our artists then plotted the results of the density readings. Using the 95% reflectance target and black and white film, a number of plots were made over the days that the filming was in progress. The range should have remained within the theoretical limits, but some of the materials fell outside of the plotted area. I do not know how many other people have experienced problems with contractors in this regard, but some portions of our film rolls had to be redone because the density did not meet the specifications on the test pattern (some were below the range, others above it). The majority remained within the range of 1.3 to 1., however.

Processing requirements for archival permanence, inspection procedures for resolution, image legibility, scale accuracy, titling, and film distortion were all evaluated in the course of inspection, as was fidelity, using a gray scale and color bar.

We used a reduction bar on this pilot study. The bar which gave the linear measure also told the reduction ratio, either 10, 8, or 5.5X. But problems occurred when the image was projected or used in a viewer. It was a clever device, but basically over-sophisticated. This over-sophistication caused it to lose its value. Making a bar that, when projected, told you the reduction ratio via its one inch scaling was too clumsy to use. The system had to be explained to each researcher and so it defeated its own purpose. We now use a metric scale in conjunction with the typical English Scale.

We are currently using three patterns on the face of the copy board. We have also changed the reduction ratios to conform to the fiche format size. This format dictated that we shoot all of our own test patterns, but with all of them showing on the test shots. We can also make the copy board itself a reflectance target of 50%. This shortcuts some of the testing procedures and allows us resolution and density tests over the full face of the chip, thereby pinpointing any distortion. Another way this helps, is that it allows us to anticipate more readily the amount of overlap necessary to capture 200 year old manuscript maps when they have to be sectionalized.

Only 13% of the maps in the pilot study required multiple images and only 6% exeeded the largest allowable dimensions of the copy board, which was roughly 37 x 59 inches. However, if analyzed closely, many of the images which were photographed separately at 10X would undoubtedly have shown improvement if microfilmed at a lower reduction ratio. The question is usually one of judgement and hinges on the legibility of the weakest lines in the original manuscript. This often forced us to drop to a lower reduction ratio and sectionalize into at least two images even with a 105mm camera.

We had trouble with the polyester jackets we used to protect the imagery. Here it was a matter of dust, which is electrostatically attracted to the surface. It appears the solution lies in the use of some type of ultrasonic sealing device. These generate a great deal of heat and must be kept at least 1/8 inch from the film image. Observing this one safeguard should preclude damage to the newly made negatives.

We finally decided to restrict the use of color. We could not obtain consistant processing even from the most sophisticated processors. Consequently, many maps which had been filmed properly had to be rephotographed and reprocessed.

We had no difficulty at all with photoprocessed maps and blueline prints. They came out very well in both color and black and white. Their legibility was tremendous. Tracing materials also photographed very well and it was only the old manuscript maps on paper which caused the real difficulties.

We have discovered two things from using the microfiche in our division. First of all, the researcher wants to have an overall view of the map record and second, the researcher wants intense magnification of a specific area of interest on the map such as a river crossing or a particular corner section of a township. Because of this, we provide both low and high magnification readers.

The Cartographic Archives Division now considers microfilming an important aspect of its work. But even at the most enterprising level, only a portion of our manuscript maps will be photographed over the next decade (perhaps on the order of 200,000 - 300,000 items). Therefore, we are forced to be selective. From the use standpoint it is very difficult to predict which records will be requested. Political activity often motivates research of a record, as has been the case recently with our material on the Panama Canal. We did not volunteer the information that we had the original French Surveys because these are in a very deteriorated state having been stored in a tropical climate for forty-five years and should be retired. Maps of this nature will not even photograph well unless restored first, a challenge in itself.

So far as the National Archives is concerned, microfilming will not mean disposal of the photographed records. This is a substitute program which will allow us to provide quick retrieval for the researcher who comes in, as well as a quick diazo copy if needed. A reader-printer works well if it has enough resolution, although the prints are not very satisfactory if they do not produce a copy as good as that of the Itek 1824, Reader/Printer, which is no longer in production. Diazo duplicates of the silver negatives, on the other hand, seem to be satisfactory while color duplicates are best accomplished by using the same Kodak SO-356 film to make a duplicate transparency as was used to make the original.

For our archival responsibilities, current color film is not adequate. When the technology improves, we will probably rephotograph many such items.

Maps and Microimaging*

Dr. John A. Wolter, Chief of the Geography and Map Division, The Library of Congress, Washington, D.C., 20540

INTRODUCTION

Most colleges and universities have map collections of one kind or another. Some of these collections are quite large and require the services of a full time staff. Others are small and are usually administered as part of one of the departments of the library. Most of the collections, whether large or small, contain valuable and ofttimes irreplaceable maps. One of the functions of the library is to preserve these maps and the rest of its collection in the best way possible. It is my feeling that provision should also be made for some type of facsimile reproduction of these maps for scholars and institutions who request it.

This paper is a discussion of ways in which the owner library can provide for the preservation, retrieval, and duplication of the rare maps of all types in its collections through the use of certain types of microimaging and reproduction equipment and processes.

Two rare maps will be used as illustrations throughout this study, although some mention of other types of cartographical materials will be made.

---

* Maps and Microimaging is a revised version of Rare Maps and Visual Aids, submitted in July 1963 to Professor Neville Pearson at the University of Minnesota, in partial fulfillment of a course and the Degree of Master of Arts in Library Science. Maps and Microimaging was presented at the 57th Annual Conference of the Special Libraries Association, Minneapolis, June 1, 1966. Neither version has been published elsewhere, and we feel this pioneering work deserves attention. Readers must understand that some of the technical data, and particularly the cost/price information, have changed, of course, in the intervening years. The presentation at Minneapolis was illustrated with eleven slides in color and black & white.

It is felt that these maps, the Westmoreland County map from the 1676 edition of John Speed's Theatre of the Empire of Great Britain and the U.S.G.S. Seattle Special 1:62,500, 1909 edition, Topographic sheet, most nearly resemble the major types of maps which must be provided for. The Westmoreland County map combines color, shading and fine engraving on high rag content paper. The Seattle Special sheet has fine engraving, uses two colors for hydrographic and topographic features and is printed on a moderately high sulfide content paper. All of the features shown on these maps should be reproduced; however, color facsimile reproduction is so prohibitively expensive that only its use in the preservation and retrieval of rare maps is dealt with.

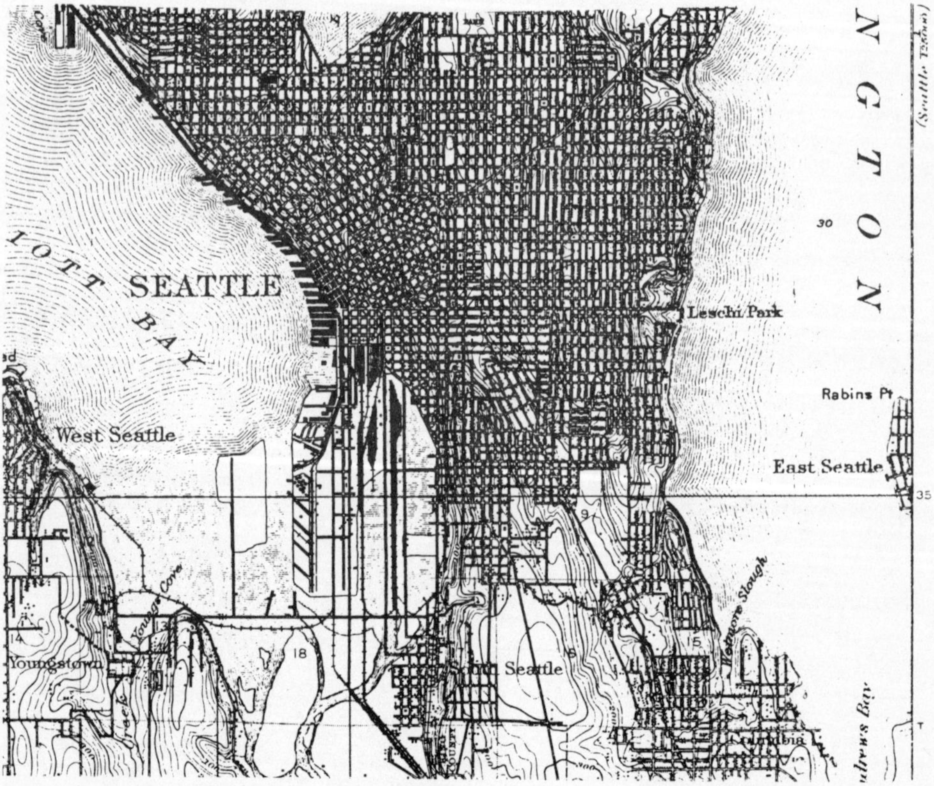

Part I

Color Slides and Prints

We are primarily concerned with the microfilming of the rare maps in our collections; however, I would like to explore, briefly, some other preservation and retrieval possibilities using filming techniques. As has been previously noted, we are dealing with two "rare" maps. Let us take one of these, the "Countie Westmorland". John Speed's Theatre of the Empire of Great Britian, is a cartographic treasure and one of the most famous atlases of all time. We are extremely fortunate in having several sheets from the 1676 edition of this atlas in our Map Library. The paper on which these maps are printed is of rag content, and very strong. The map sheets have survived, probably as a bound atlas, for almost three hundred years. Lately, however, some wear has become apparent as the maps are handled by persons using them for study. It was felt that we should preserve the maps in the best possible way, but within the limitations imposed by our budget. One can always back rare maps with a number of different materials, but this does not always decrease the possibility of handling damage or of damage by fire, insects, or other destructive forms.

In order to help to preserve these maps I felt that they should be microfilmed. However, I also realized that even though they were microfilmed it would not be possible to reproduce them in their original size and color without a greater expense than my budget could stand. The user of the map would, I am sure, prefer a faithful color reproduction if it was available. (Providing, of course, that the original map was in color). I therefore photographed the maps on 35 mm color film and made 2 x 2 in. film slides.

These color film slides can serve many useful purposes. It is possible to photograph a catalog card at the same time as the map, but unless it is projected at the same scale as it is photographed, it is quite difficult to read.

The color slide can be attached to the catalog card (see Exhibit 1). It is assumed that the rare maps have their own catalog, and that there should be no trouble in keeping the two objects together. If the maps do not have their own catalog, the slides and the catalog cards can be numbered identically, probably by classification number, and stored separately. With the color slides one can either project the image or view it in an inexpensive hand viewer. Unless the map library can afford a self-contained slide projector of some type, the latter technique must be used.

The map librarian should, of course, have color slides of the cartographic treasures in his collection, and it is assumed that most libraries can afford this expense. He should also be able to provide color slides to professors who use, or could use, rare maps to illustrate their lectures. The use of color slides of maps in courses in exploration and discovery, history, and political science, in addition to their use in the usual cartography and map interpretation courses, would certainly enhance the value of many lectures.

With color prints it is possible to carry the techniques of preservation and retrieval one step further. Exhibit 2 shows a 3½ x 5 in. enlargement of the 35 mm color slide. The verso of this enlargement could contain the information one would normally find on a catalog card. The enlargement serves a double purpose. First, it is a photographic record of the cataloged map and can be read with a magnifying glass for detailed study. Second, it is a catalog card. The card can be bonded in clear plastic and thus preserved for a long period of time. A larger card, of 5 x 7 inches, represents a further refinement of this technique. The inclusion of more information and easier facility in reading, make this card the optimum enlargement size. As can be seen in Table 1, the cost of this latter size might be prohibitive. As a matter of fact, the cost of the smaller size enlargement might be prohibitive also, but

Exhibit 1. Color slide, catalog card and shelf list card.

212:p:1676

Area: Westmoreland Co. England Location:

Title and edition: The countie Westmorland and Kendale the chief towne described, with the Armes of such Nobles . . .

Author: Speed, John

Place, publisher and date: [London], author, [1676]

Size: 15"x20" Scale: 2 1/2 Eng.mi. to an in

Type (check): Political X Physical X Historical X Economic

Condition (check): Mounted Unmounted X Folded Rollers

X Notes on back

Accession no.: Ref. 17

5M—3-55

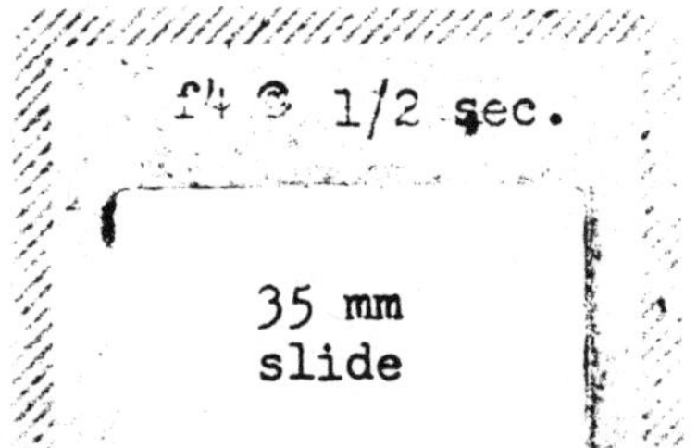

212:p:1676

SPEED, JOHN, 1552-1629

The countie Westmorland and Kendale the chief towne described, with the Armes of such Nobles as have been Earles of either of them. [London, 1676]

15 x 20 in. (From Speed, John; Theatre of the Empire of Great Britain...London, 1676.)

Scale: 2 1/2 English miles to an inch.

Insets: Kendale; [arms of] Iohn D. of Bedford, Iohn D. of Somerset, Iohn de Foix.

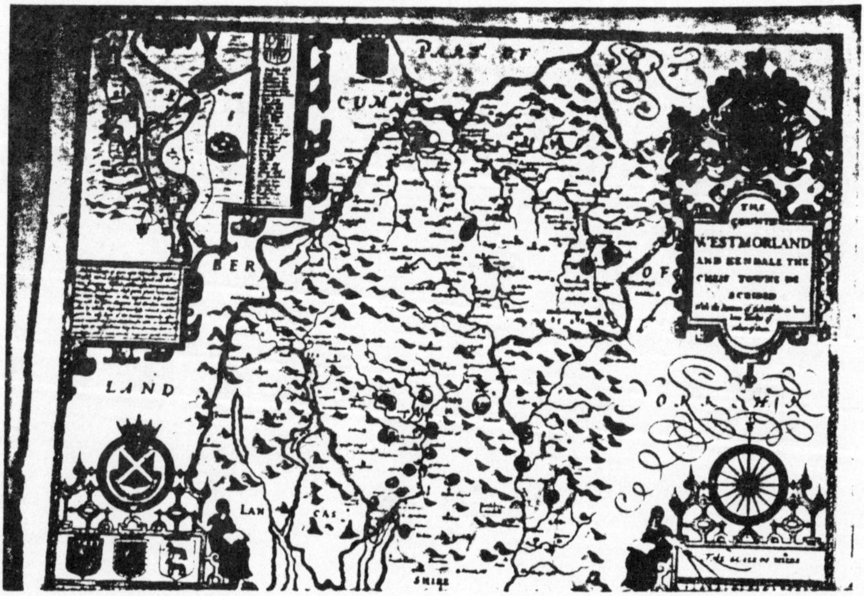

SPEED, JOHN, 1552-1629

The countie Westmorland and Kendale the chief towne described, with the Armes of such Nobles as have been Earles of either of them. [London, 1676]

15x20 in. (From Speed, John; Theatre of the Empire of Great Britain...London, 1676.)

Scale: 2 1/2 English miles to an inch.

"Performed [by] Iohn Speed and are to be sold by Thomas Bassett in Fleetstreet and Ric, Chiswell in St. Pauls Churchyard."

Insets: Kendale: [arms of] Iohn D.of Bedford, Iohn D.of Somerset, Iohn de Foix.

Exhibit 2. Color print catalog cárd - 3 1/2" x 5".

Table 1 Map Mounting and 35 mm Color Filming Costs

a. Map Mounting[1]

Median size for cost purposes (18" x 36". $4.00)
All sizes under - cost by time required.
All sizes over - $.25 per square inch over base.
"Countie Westmorland" . . . . (15" x 20") $2.50

b. 35 mm Color Filming[2]

Roll film[3] (including developing)			
Kodacolor	CX 135 - 20	(negative)	$2.45
Kodachrome	KR 135 - 20	(positive)	$3.20
Kodachrome	KX 135 - 20	(positive)	$3.20
Ectachrome	EH 135 - 20	(positive)	$3.70
Printing (per print)			
Kodacolor	CX 135 - 20	(3-1/2" x 5")	$ .30
Kodacolor	CX 135 - 20	(5" x 7")	$1.25

c. Total Cost of Color Film Slides

Least expensive	KR 135 - 20	(per slide)	$ .16
Most expensive	EH 135 - 20	(per slide)	$ .19

It should be noted that printing from color film slides is more expensive than printing from the negative color film and a good deal less satisfactory.

---

[1] University of Minnesota Library Bindery Charges. The mounting, if done by the map librarian, would cost approximately the same amount.

[2] Commercial photographer quotes. University Photography Laboratory costs are slightly more. The actual photography can be done by the map librarian if he has the proper equipment and the technical competence.

[3] Film speed and the cost of the film increase together.

the advantages of having everything appear on one card seem to outweigh this factor.

The color slide and print techniques of preservation and retrieval are both worth serious consideration for use as permanent records of the rare materials in a map collection. In addition to the features which might have been previously discussed, these techniques preserve the color of the original map and this fact alone makes them important. I am sure that there are some libraries which are well able to afford the cost of these techniques and even some libraries which could afford to microfilm as well.

Microfilming has definite advantages over all of the techniques discussed above. The chief drawback of microfilming is the impossibility, at the present time, of cheap color reproduction. It is by belief that the advantages of microfilm (black and white) far outweigh this one advantage, particularly for those materials which are most easily preserved in this way.

Part II

Microfilm and Hard Copy

It has become apparent to most educational institutions in the last few years that microfilm in one form or another, is a valuable research tool. Microfilm cost is relatively low, it requires much less storage space than ordinary library materials, and it can be projected or viewed in a microfilm reader or by a hand viewer. In short, it is quite a significant advance in communication technology.

Microfilming is defined by Boucher in his Fundamentals of Photography, as "The reproduction of subjects on small negatives and positives in which fine detail is too small to read with the unaided eye."[1] Scanlan describes microfilm as follows:

> *Microfilm consists of an acetate base with one side coated with a silver halide emulsion. This emulsion is designed for recording fine line work with a minimum of distortion of toning. (This is normally referred to as first generation film.) When processed, the film is in a negative form. Additional copies can be made by contact printing on second generation print film. This can be done by diazo or Kalvar methods. The print will be positive if the original is positive, and negative if the original is negative. One can also obtain Kalvar positives from diazo negatives. Additional copies can be printed from the second generation print film.*[2]

I will not discuss the exclusive use of roll microfilm in this paper even though Parker[3] has concluded that it is cheaper. The difficulty of cataloging the maps on roll microfilm and the probability of scratching and otherwise disfiguring it when using the reader printer, preclude its use. A scratch on the microfilm of a map may be seriously misinterpreted when it is viewed or printed. I have limited myself to the use of the diazo and Kalvar aperture cards, which are subject to the same possibility of damage, but to a much lesser degree, and with much less disastrous consequences.

Aperture cards were born of wartime necessity when the flood of material used for intelligence purposes became too great to contend with by using the older and slower systems of filing and recording; however, it was not until 1958 that the microfilm industry produced a reader-printer specifically designed to enlarge filmed images and to produce dry copies. With both aperture card and reader-printers the use of microfilm has increased enormously.

Microfilm has two distinct advantages in addition to the advantage of preservation. It is easy to handle and it requires little space. In some places, such as on shipboard, the space saved could be of great significance. It should be noted that a 22-drawer tab filing cabinet can hold over 83,000 aperture cards. There are two disadvantages. A reader is required, and some means of reproduction is needed.

There are many problems in map microfilming. The microfilm must have at least the same standards of accuracy that the original map has. Wertz states in an article in National Micro-News, that reproduction should be in color and the same size as the original, and that the reproduction system must be portable.[4] He is speaking here of an ideal system, and one for standard government maps of uniform size. The U.S. Geological Survey 1:24,000 or 7½-minute quadrangle reduced 20 times, will fit a 35 mm frame. The map itself has a maximum of 1/50 of an inch distortion. The camera lens should have the same maximum. The projector must have an optical system at least as good. Wertz further states that the major obstacle is the faithful reproduction of color.[5] It is important to note here, however, that color film is only capable of one-third the resolution of black and white microfilm. Incidentally, the "AHU" master microfilm that was used in this study was checked by the National Bureau of Standards resolution charts at 158 lines per millimeter. Excellent resolution when you consider that the government standard is around 118 lines per millimeter.

Except for this brief mention, it is not possible to examine the problems of modern topographic map color filming and stay within the scope of this paper. This may very well be the subject of a future study. I am primarily concerned with our two "rare" maps and how they can be handled with the technical equipment that was made available to me.

Equipment and Processes

The microfilm and processing equipment used was provided by the Diabetes Documentation Project, Anatomy Department, University of Minnesota, and by the Minnesota Mining and Manufacturing Company. Table 2 illustrates the cost of the microfilm equipment and accessories. It is felt that most college and university libraries could use a reader-printer of the "Filmac 200R" type and

Table 2 Microfilm Equipment and Accessories Costs

a. Microfilm Equipment Costs

Recordak MRC-4 camera	\$5,000.00
"Filmac 200R" reader-printer[1]	\$1,685.00
"Uniprinter" copier 086 "D" aperture.	\$1,021.00
"Roll to card printer" estimate[2]. . .	\$1,000.00
Total fixed cost. . . .	\$8,706.00

b. Accessories Cost

"Thermo-fax" microfilm printer activator (pint).	\$ 1.85
"086" Ammonia developer (quart) . . .	\$ 1.55
Recordak "AHU" 35 mm film-220 (roll).	\$ 8.50

[1]Includes roll film accessory.

[2]None available commercially as yet which work correctly. The Diabetes Documentation Project roll to card printer (16 mm) works very well. With modification it could be used for 35 mm microfilm.

size for a variety of jobs. The engineering and other technical libraries of the college or university could share both the cost of and the use of this equipment with the map library.

The "Filmac 200R", which was used for most of my reproduction, is a reader-printer unit consisting of a metal housing containing an illumination system, an optical system, a viewing screen, a film-transport system, a system for both holding and transporting the sensitized material, and a processing section for electrolytic development of the latent image. The intensity of the current controlling the deposition of the image and the exposure duration are controlled by knobs located on the front of the machine.

Hawken explains the electrolytic process in the following manner:

> *In this electrolytic process, the sensitized material consists of three layers: a paper base, a thin layer of a metal foil such as aluminum, and, on top of this, a coating of zinc oxide as a resin binder. This zinc oxide coating provides a photo conductive layer which in the dark acts as an insulator; i.e., it has high electrical resistance, the image areas having now become electrically conductive through the action of light. If the exposed sheet is then placed very briefly in contact with a suitable electroplating solution and a d.c. electric potential applied, current can flow between the solution and the aluminum-foil layer. This causes metal ions in the solution to deposit on 'plate out' on the zinc oxide surface in the image areas to form a visable image.*[6]

To use the Filmac 200R as a reader-printer, three things are necessary: 1) a roll of "Filmac" microfilm copy paper, 2) "Thermofax" microfilm print activator which processes the exposed copy paper, and 3) the microfilm which is to be printed.[7]

Prints are supplied, depending on the exposure setting, in from 2 to 31½ seconds. The former figure denotes a "light" setting and the latter a "dark" setting.

The microfilm roll or aperture card is inserted in its proper place on the front of the machine. The microfilm is projected on the screen, positioned properly, and focused by means of control knobs. To make a print the timer knob is set for the desired exposure interval and the print button is pressed. When the exposure is completed the finished print (hard copy) is advanced out of the right side of the machine. It can then be torn off against the cut-off bar and used immediately.

I should say a word now about a new development, the 3M "333" Dry Silver Printer which will print, on low cost paper (about $.09¢ per copy), an 18" x 24" image. This is a quite recent development, shown for the first time at the National Microfilm Association Convention last week. I have not seen it although I have tested the 3M "222" Dry Silver Printer and found it to be very good. Further comparisons must be made before a value judgement can be made.

I have not confined myself exclusively to 3M products however. I have used and evaluated the Recordak 1824 model reader-printer, the Xerox 1824 printer, and the Itek 1824 reader-printer with the same type master film. So far none of them have given me as good resolution as the "200R" has. It is entirely possible that the Recordak 1824 or the 3M "333" will both eventually prove to be better than the "200R".

The equipment used in the study, in addition to the Recordak camera, included the Filmac "200R" microfilm reader-printer, the "Uniprinter" copier 086 "D" aperture card duplicator and a Kalvar heat developing unit. The Taylor -Merchant hand viewer was useful for inspecting the color film microfilm and aperture cards.

Method

The National Bureau of Standards resolution charts were filmed at all of the densities available with the MRC-4 camera, and at reduction ratios of 12:1, 15:1, and 18:1. The maps were microfilmed at the same densities and at the proper reduction ratio. The film was developed and the resolution charts checked by microscope. At an 18:1 reduction ratio the resolution was 158 lines per millimeter in the center and 135-148 lines per millimeter on the edges of the microfilm frame.

The master map frames were checked in the "200R" microfilm reader-printer for readability, printed out and the hard copy checked by jewelers glass for flaws. The best density master frames were chosen and reproduced in second generation diazo aperture cards on the "086 D". Hard copy from the second generation aperture card was made and checked. Third generation diazo aperture cards were made and also checked. All of these diazo aperture cards were made for viewing and/or printing. From the second generation diazo aperture card I also made a Kalvar positive aperture card.

Imagine yourself as an instructor teaching a course in urban geography, city planning or map interpretation. You have been able to obtain the master aperture card from the map library in your school or department or perhaps from a regional depository. You took your master card or cards and had your department secretary make positive cards for your use in class. You can show this aperture card image by use of any one of several new aperture card projectors. We have used several and made some recommendations for modification. There are technical problems with these new devices but nothing that can't eventually be solved. I personally think that a projector of this type has very great potential for effective teaching. It makes use of a relatively inexpensive technique for course enhancement and enrichment.

Let us get back to the diazo film and the reader-printer. The flow chart (Exhibit 3) illustrates the reproduction process.

The hard copy, produced from the "200R", is legible, the contrast is good, and the copy strong enough to withstand fairly rough usage. It can be written on with very soft pencil or crayon. The cost of reproducing hard copy is low enough--24 cents apiece (Table 3)--to encourage the use of several copies of the same map. It would be less expensive to do this than to make <u>all</u> of one's

Exhibit 3 - Flow Chart

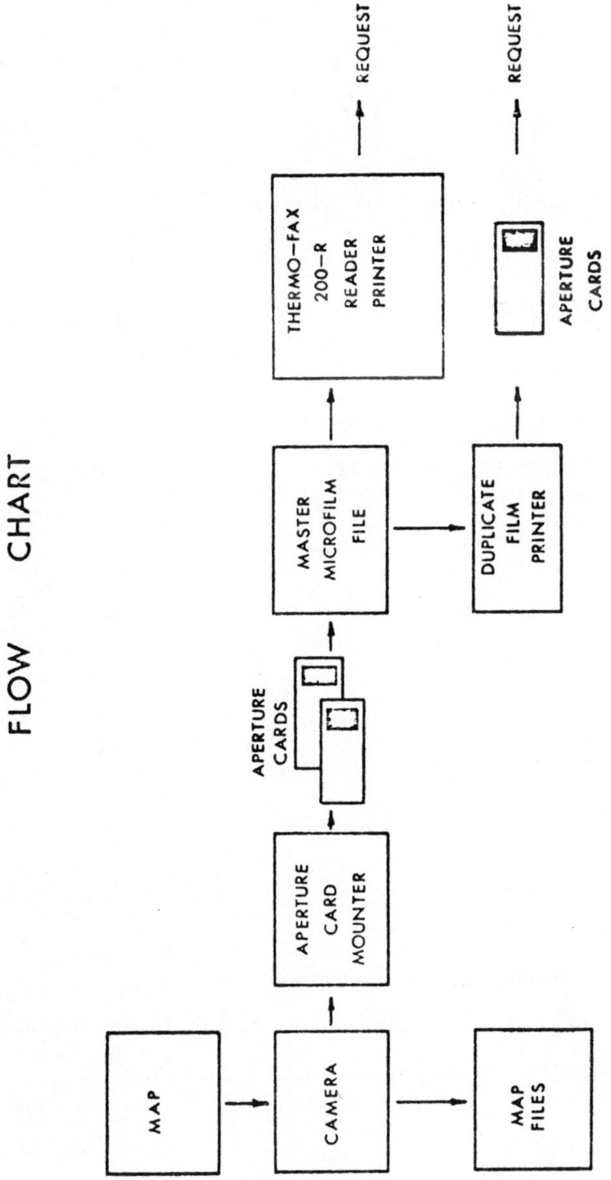

Table 3 35 mm Microfilming Costs[1]

35 mm "AHU" microfilm, 100 ft. roll (including processing).	$ 8.15
35 mm "AHU" microfilm cost per exposure	$ .02
"Filmsort" aperture card "duplicard D"[2]	$ .03
"Filmac" Microfilm Copy Paper (18" X 24")[3]	$ .24
The total cost of providing a "Filmsort" aperture card "D" Single is	$ .05
The total cost of providing a "Duplicard" is	$ .08
The total cost of providing an 18" X 24" copy is . .	$ .32

---

[1]Costs are rounded off to the nearest cent.

[2]"Duplicards" 5" X 3-1/4" and priced at $ 44.11 per thousand

[3]18" width roll 237 feet in length at $ 22.88 per roll (100 rolls.)

notations on one very expensive photographic reproduction with the attendant possibility of error.

Hawken has compared several reader-printers in their relation to library usefulness and has found most of them wanting in one respect or another.[8] He does not discuss map reproduction. He notes that the "Filmac 200R" was designed primarily for the reproduction from microfilm of blueprints and engineering drawings.

It can be seen from the present study that this machine is also very useful for the reproduction, from microfilm, of rare maps.

Librarians should carefully note Hawken's recommendations and conclusions. According to his criteria the "Filmac 300" is the only really useful reader-printer for all-around library use that is manufactured today. I am not expert enough to dispute this statement, but the "Filmac 300", as good as it is, does not produce large enough hard copy for good map facsimile reproduction. As previously stated, there are new reader-printers on the market that may be better than the one that I found best, the "200R". One of the problems with any research in this area is the profusion of devices of all types--new models and modified older models--which are constantly being unveiled. Because of this, I think that we should tell the manufacturer <u>what we want</u> and let him develop a machine or machines which will do the job for us and not have to adapt to available equipment, none of which is really what we need.

Part III

Summary and Conclusion

The major differences between the two techniques described are: color vs. black and white, speed of retrieval and duplication, and cost.

The 35 mm color slide is not too expensive, and the cost to the library of recording all of its rare maps in this manner would probably not be too high. The cost of printing, in color, of any size over a few square inches would be prohibitive. As previously discussed, the slides are useful for other purposes and they can be reproduced (as slides) at a nominal cost. (30¢ each)

Microfilming of the rare maps is less expensive and the microfilm, reproduced in master aperture cards, is much more easily and cheaply reproduced; however, the equipment required for the entire microfilm reproduction process is relatively expensive. If this equipment can be used by other departments within the library system, its purchase is highly recommended.

It has been seen that the color slide and print techniques lend themselves to the efficient retrieval of information. They are at least as efficient as the normal cataloging system, and perhaps more so because the map information can readily be seen with the aid of a hand viewer. In most cases it would still be necessary to use the original map, but the slide and print do help a good deal in the selection of maps to be studied.

The aperture card was designed for the quick efficient machine retrieval of information. The card can be punched in many different ways representing as many different subjects. Much more information can be brought out on the aperature card than can be noted on a normal catalog card. I have not discussed this aspect of the retrieval of information because it has been well explained by Mr. Easton in his paper and by Mrs. Logan and Mr. Palmerlee. If the entire map collection, in my case over 150,000 maps, but in any case at least several thousand maps, were put on microfilm the expense of the proper machinery might be justified. With the modern large scale topographic map, both foreign and domestic, selling for about $1.00 and in most cases no more than $.50¢ per sheet, it would seem that the microfilming of these maps would not be needed for preservation purposes--however, they should be available as slides. Those maps, which are difficult to obtain, e.g. topographic maps from the USSR and Eastern Europe, and those maps which by their age or condition have become rare should be microfilmed. Important examples of the latter type of maps are the United States Geological Survey Topographic maps, similar to the Seattle Special, which were printed on cheap wood pulp paper around the turn of the century, and which have now become so brittle that a slight touch will cause them to disintegrate. These topographic maps must be preserved. I am sure that copies are preserved by the USGS, but to my knowledge no microphotographic record capable of duplication has been made.

The aperture card, as used in this study, seems to be an efficient way in which to preserve, retrieve and duplicate the rare maps in the map library. The master aperture card can be duplicated and mailed to the scholar or institution requesting such information for approximately 14-cents. A hard copy close to or the same as the original map size, up to a maximum of 18" x 24", can be produced for less than 25-cents. It is my belief that this technique will prove to be a most efficient and useful way of providing the patron of the rare map collection with what heretofore has been almost impossible; an inexpensive large copy of the rare map to be studied.

Appendix

DIAZO-TYPE FILMS

General Film Characteristics

Diazo microfilms have come to be used widely as copy films because of their ease in use, because they are nonreversal and because of their high resolution capability.

Diazo is relatively new to the reproduction field. It was discovered in Germany as the result of research with diazonium salts (also called diazos of azo compunds). Certain diazonium salts can be transformed into a colorless compound by ultraviolet light. The remaining diazonium salts can be converted to an azo dye by reaction, under alkaline conditions, with chemicals called couplers. The diazonium salts are mixed with couplers and acid stabilizers to form the light-sensitive coating. The coating is applied to an acetate base. When exposed to ultraviolet light through a microfilm, the line work, being of low density, permits the light to go through and decompose the diazonium salt. During the development, in the presence of ammonia gas for alkalization, the portion which was not exposed becomes colored when the coupler reacts with the diazonium salt. Hence the line work is clear and the background is a dark stabilized dye. The name "diazo" is applied since the azo compound contains two nitrogen atoms (from the Greek: di, two, and azote, nitrogen).

Kalvar Image-formation Principle

Kalvar film produces an image in a radically different way than silver emulsion or diazo films. The process is described briefly by Dr. Nieset, Director of Research at Kalvar Corporation, as follows:

> *In the Kalvar film process image formation is accomplished by light scattering rather than light absorption. A compound which will undergo photolysis to yield volatile products is dispersed in a polymeric vehicle which can be relaxed by heat. Upon exposure of light, internal pressures are produced within the vehicle resulting in a 'latent image' of internal stresses. Upon subsequent exposure of heat, reorientation and an ordered recrystallization of the polymer results from the force of these internal pressures. The reoriented and the recrystallized portions of the polymer scatter light strongly to produce the developed photographic image.*

Density Produced by Light Scattering

When light strikes silver halide crystals it is absorbed, but in Kalvar film it is scattered by scattering centers. The light transmitted by Kalvar film is reduced in proportion to the light lost in scattering. Only a small portion of the incident light is absorbed, while a major portion is scattered. The eye sees only that portion which falls upon it. Areas from which light has been scattered away from the eye are interpreted as dark areas. Light absorption plays the fundamental role in conventional silver photography; light scattering plays the fundamental role in Kalvar. (from: Nelson, Carl E., Microfilm Technology; Engineering and Related Fields. New York, McGraw-Hill, 1965.)

Footnotes

[1] Boucher, Paul E. *Fundamentals of Photography*. Third edition. New York, Van Nostrand, 1955. p. 366.

[2] Scanlan, James R. *The Active Uses of Microfilm*. New York, Wolf Business Publications, 1962. p. 117.

[3] Parker, F. M. "Engineering drawing processing system." *Special Libraries* 51:429:432. October, 1960.

[4] Wertz, John A. "A System for Microfilming Maps." *National Micro-News*. No. 54, October, 1961. p. 117.

[5] *Ibid.*, p. 119.

[6] Hawken, William R. *Enlarged Prints from Library Microforms*. Chicago, American Library Association, 1963. p. 10.

[7] *Ibid.*, p. 31.

[8] *Ibid.*

The Appendix is from: Nelson, Carl E. *Microfilm Technology; Engineering and Related Fields*. New York, McGraw-Hill Book Company, 1965. pp. 185; 202-203.

Annotated Bibliography

Larry Cruse, Head, Map Section, University Library, University of California, San Diego. La Jolla, CA 92093

1. Aldrich, Robert C., James von Mosch, and Wallace Greentree. 1972. Projection-viewer for microscale aerial photography. Berkeley, CA: U.S.D.A. Forest Service, Pacific Southwest Forest and Range Experiment Station. USDA Forest Service Research Note PSW-277.

Homemade 70mm projector project illustrating the fundamentals of projector design and construction. Includes zoom capability and gridding of the projected image. Well illustrated.

2. American Geographical Society. [1977?] [Microfilm and microfiche of selected rare atlases in the Society's collection.] New York: author.

Interrupted by the AGS Map Collection's move to the University of Wisconsin, Milwaukee, this pioneering program in preservation and distribution will probably be revived in the near future.

3. Amidon, Elliot L., and Joyce E. Dye. 1976. Microcopying wildland maps for distribution and scanner digitizing. Berkeley, CA: U.S.D.A. Forest Service, Pacific Southwest Forest and Range Experiment Station. USDA Forest Service Research Paper PSW-114/1976. (NMA 76-0507)

Essential reading on all aspects of microcartography, including computer input microfilm (CIM), film types, resolution factors and color versus black-and-white, it is the best primer on the subject.

4. Anderson, W.H., and F.X. Wallner. 1978. Small format aerial photography: a selected bibliography. Sioux Falls, SD: EROS Data Center. (NTIS PB 279 849)

Bibliography of literature in the EDC library on 35 to 70mm imagery; 21 annotated citations.

5. Andrews, Peter. 1977. National Air Photo Library. Micro Notes 5(3):16-17.

Pioneering use of "browse files" to graphically reference remote sensing.

6. Approach. 1974. Advanced aircrew displays. Approach August 1974:1-3.

View of a modern cockpit with integral microform map display.

7. Archard, T.N.J. 1968. Microfilm standards. In: National Reprographic Centre for Documentation, A technical appraisal of the 70mm format for map reproduction, pp. 43-44. Hatfield: Hertfordshire County Council, Technical Library Service.

General introduction to the need for, and evolution of, standards, with emphasis on British experience with map microforms.

8. Aron, Albert. 1970. Scientific uses of computer graphics equipment, (from a slide presentation). In, National Microfilm Association Proceedings XIX: 232-246.

Covers remote sensing formats from $14mm^2$ to $45mm^2$, pointing out the information density (1,000,000 addressable points).

9. Babcock, Homer C. 1978. Automated cartography data formats and graphics: the ETL experience, The American Cartographer 5(1):21-29.

The U.S. Army Engineer Topographic Laboratories (ETL) probably has more experience with the high technology generation and application of cartographic microforms than any other research body. This is a summary of their experience, and, by implication, the state-of-the-art for integrated technologies under "real world" conditions.

10. Balasubramanian, N., and R.D. Leighty. 1974. Coherent optics in mapping. Society of Photo-optical Instrumentation Engineers Proceedings 45:187-191.

State-of-the-art as of 1974.

11. Bartik, B., and Z. Hanus. 1975. Mikrosnimkovani map velkych meritek. Geodesia es Kartografia 21(10):285-291

Microfilming of large-scale maps. In Czechoslovakian.

12. Bell, S.B.M., and Woodsford, P.A. 1977. Use of the HRD-1 laser display for automated cartography. The Cartographic Journal 14(2): 128-134.

Use of computer driven laser in the production of printing masters on film. Bibliography.

3. Berry, H.A., and P. Horowitz. Investigations of the tactical interpretability of topographic displays. In: Report From the Seventh Annual Army Human Factors Engineering Conference, pp. 121-126. Washington, D.C.: U.S. Army, Office of the Chief of Research and Development.

Human factors assessment of aircraft Projected Map Displays (PMD). Points out the fundamental importance of color in critical use situations, as well as tonality and symbolization, all of which will have their impact on general map design, but especially on map microforms.

4. Best, Robert G., and Jack R. Smith. 1978. Photographic contrast enhancement of Landsat imagery. *Photogrammetric Engineering and Remote Sensing* 44(8): 1023-1026.

"...varying film type, developer, and development time to produce higher contrast in the reprocessed image," using simple photographic procedures. Exploits the differences between the 64 grey tones in a 70mm Landsat MSS scene. Yields the same results as computer enhancement at a fraction of the cost.

5. Best, Thomas D. 1965. 35-mm slides of topographic maps: pitfalls and prospects. *The Professional Geographer* 17(3):20-23.

16. Bier, Robert A., Jr. 1978. On-demand computer cartography and its effect on map libraries. *Special Libraries* 69(2):61-65.

Computer Output Microfilm is treated on p. 61. Figure 4 reproduces a computer generated map microform. Notes the need for a centralized registry of cartographic data bases, and predicts the probable future role of the map librarian in the information chain.

17. Blessing, Don. 1972. General aerial photography. In: Hasselblad, *Aerial Photography*, pp. 5-10. Goteborg, Sweden: Victor Hasselblad Aktiebolag. (Available from Braun North America.)

Current capabilities of "medium format" 70mm cameras.

18. Bohem, Hilda. 1976. A visible file catalog for photographic materials. *The American Archivist* 39(2):165-166.

Equally applicable to map collections, this discussion of current practice at UCLA points out the multiple advantages of including a microprint of the cataloged material on the catalog card.

19. Bogo, Laszlo, and Mihaly Marosi. 1973. A foldmeresi adat- es terkeptarak mikrofilmestitesenek helyzete Magyarorszagon. Geodesia es Kartografia 1973(4):268-273.

Use of microfilm as a map preservation medium in the Hungarian archives. In Hungarian with Russian

20. Borgstede, Roy F. 1977. Color statistical maps. Journal of Micrographics 11(1):46-48. National Micrographics Association Proceedings 26(Part II).

U.S. Bureau of the Census Computer Output Microfilm (COM) color seperation program. Output size is 310mm (8 1/2 x 11 inches).

21. Bradford, Robert S. 1973. 70mm electron beam imagery. Society of Photo-optical Instrumentation Engineers Proceedings 32:129-136.

Description of the ERTS A Electron Beam Recorder (EBR), the 70mm digital to analog interface between the satellite and the graphic image.

22. Brandli, Henry W. 1978. The night eye in the sky. Photogrammetric Engineering and Remote Sensing 44(4):503-505.

Visible and near-infrared space imagery, available in 35mm format, in addition to others.

23. British Standards Institute. 1977. British Standard 5525, specification for 35mm microcopying of maps and plans. London: author.

24. Broome, Frederick R. 1974. Micrographics: a new approach to cartography at the Census Bureau. American Congress on Surveying and Mapping, Proceedings of the Fall Convention, Washington, D.C., September 10-13, 1974:1-13.

see: Borgstede, 1977.

25. Bruno, Michael. 1943. Maps on microfilm, some factors affecting resolution. Journal of the Society of Motion Picture Engineers 41:412-425.

While microreduction may be an invention of the 19th Century, modern microfilm technique is a by-product of motion picture hardware. This early account is concerned with critical elements in the system train, especially the lens and film, both of which have been greatly improved.

6. Buchan, B. 1973. Microfilming maps, architectural and constructional drawings. Delivered at the World Micrographics Conference. (NMA 73-0238)

7. Cameron, George. 1951. Applications of 70mm microfilm. National Microfilm Association Proceedings 1:39.

The economics of 70mm microfilm and brief mention of its potential for remote sensing and mapping applications.

8. Cann, Edward H. 1958. Microfilming in color. National Microfilm Association Proceedings 7:7-13.

Color maps (p. 10), their use by the City of New York (p. 11), and Saint Louis University's project to microfilm the Vatican Library manuscripts, an unknown number of which are cited in Roberto Almagia's Monumenta cartographica vaticana...., V. 1. Planisferi, carte nautiche e affini dal secolo XIV al XVII....

0. Carneggie, David M. Large scale 70mm photography - a potential tool for improving range resource inventories. American Society of Photogrammetry. Papers from the 34th annual meeting: 79-90.

1. Carr, William H[oward]. 1962. Microfilming in mapping applications. National Microfilm Association Proceedings 11:249-254. (NMA 62-0021)

Map microfilming experience of the Graphics Division, U.S. Army Geodesy, Intelligence and Mapping Research and Development Agency, particularly 70mm color separations for full-scale color map regeneration.

2. Carr, William Howard. 1967. Military micromap display. The Military Engineer 59(389):180-181.

3. Carr, W[illia]m Howard. 1976. Electro-optical systems and requirements in cartography. Proceedings of the Technical Program, Electro-Optical Systems Design Conference [and] International Laser Exposition, New York, NY, Sept. 14, 15, 16, 1976: 339-343.

Summary of U.S. Army Engineer Topographic Laboratories (ETL) experiments with microimagery, Electron Beam Recorders (EBR), and lasers in cartographic production via computer output.

34. Carroll, C. Edward. 1978. Bibliographic control of microforms: where do we go from here? Microform Review 7(6):321-326.

Summarizes the current problem confronting libraries which own microform sets: they are not analyzed (the individual works are not separately cataloged, ergo, they are "invisible" bibliographically). Extensive notes documenting literature on this subject. Of interest to map libraries because there is no systematic means of knowing which cartographic products are in these sets, which often leads to duplicate purchases. Essential reading for anyone interested in pre-twentieth century geography/cartography.

35. Cerwin, F.A. 1960. Microfilm repros stream-line Chicago Highway Department. Industrial Photography 1

36. Clare, W.G., and A.G. Sabin. 1968. Mechanical accuracies required for a 70mm microcopying camera, In: National Reprographic Centre for Documentation, A technical appraisal of the 70mm format for map reproduction, pp. 31-42. Hatfield: Hertfordshire County Council, Technical Library Service.

37. Clark, Robert K. 1976. Graphic COM and CIM systems: computer graphics at a large IBM installation. National Micrographic Association Proceedings 25: 328-331

Includes mapping functions as integral input/output functions at the Argonne National Laboratories.

38. Clegy, R.H., and J.P. Scherz. 1975. A comparison of 90-inch 70mm and 35mm cameras. Photogrammetric Engineering and Remote Sensing 41(12):1487-1500.

Cost, resolution and photo interpretation potential comparisons for color and color infrared aerial photography. Points out the improvements in lenses and films.

39. Coates, Bryan E., and Alan M. Hay. 1974. South Yorkshire County is born and analyzed. The Geographical Magazine 56(8):395.

COM plotting on photo film using 16 levels of tonal intensity.

40. Cohn, Dr. Morris M. 1960. Municipal arteriosclerosis and hardening of the records. National Microfilm Association Proceedings 9:188-192.

General recommendations for the use of microfilm by local governments, including such mapping applications as underground utilities and general mapping.

. Cooper, G.H. 1978. Half tone photography for thematic maps. Cartography 10(3): 150-153.

A technique for which map microfilms would make an excellent base.

. Cornwall, Bruce, and Arthur H. Robinson. 1966. Possibilities for computer animated films in cartography. Cartographic Journal 3(2):79-82.

Kinetic cartography using microfilm as the output medium. Because of the large computer store required, these are still relatively simple in their progressions.

3. Crespo, Carmen. 1975. The Spanish Microfilm Service. International Council on Archives. Microfilm Committee. Bulletin 4:49-

Have an extensive map microfilm inventory but are disappointed in the lack of sales, especially to Western Hemisphere clients.

4. Critchley, W.E. 1970. Old maps? Photocopying helps in using them. Microdoc 9(3):68-69.

Use of 35mm for reference and 70mm for preservation of valuable maps at the Aberdeen City Library.

5. Cruse, Larry. 1979. Collecting microcartography: sources and prospects. Paper delivered at Special Libraries Association, Geography and Map Division, June 11, 1979, Honolulu, Hawaii.

6. Cunningham, M.O. 1969. War-time microfilming activities, OSS China Theatre 1943-1955. Journal of Micrographics 3(1):14-17.

History of the most extensive map microfilming project ever undertaken, including the major map collections of the United States and Great Britain. Unfortunately, the films are now conspicuously absent from the holdings of the U.S. National Archives and are unlocatable. See also, annotation on entry number 229.

47. Dimiani, Byron. 1970. Microfilm in a title company. National Microfilm Association Proceedings 19:131-134.

Includes discussion of land records associated with real estate transactions.

48. De Ferry, Ferreol. Ulane Bonnel, translator. 1969. Photoduplication at the Archives de France: practice and policy. News from the Center 6:5-11.

Brief mention of map microfilming program. Explains role of the Societe Francaise du Microfilm, a commercial supplier operating within the French Archives.

49. Delehanty, E.J. 1959. Map microfilming reduces storage problems. Public Works 90:127-128.

Local government records and map microfilming by Westchester County, New York.

50. Devine, E.P. 1972. System M: a proposed method for producing aeronautical charts using Computer Output Microfilm. Presented at the International Cartographic Association Conference, Ottawa, Canada, August 1972.

For a published discussion of this work, see: Grosso, 1977. The object is to computerize frequently revised charts for rapid updating and distribution.

51. Dorfman, Harold H. 1974. Color microfilm program (City of New York). National Microfilm Association Proceedings 23(II):303-305. (NMA 74-0070)

Includes discussion of city land map preservation and updating program. Maps are 35mm, formatted 12 per 105mm microfiche. Cold storage is used for micro preservation.

52. Dorfman, Harold H. 1976. Color microfilm records. The Office 83(4): 18-19, 189. (NMA 76-0237)

Update of the New York City program.

53. Drews, H.E. 1969. National Park Service's photo-drafting tool - 105mm. Engineering Graphics 9(9):30-34,46.

54. Eastman Kodak Company. Maps on microfilm are carried in Indianapolis Water Company service trucks. The Kodak Compass 1973(4):1-3.

Maps of the entire water pipe network are carried on each truck and are inexpensively and quickly updated.

55. Eastman Kodak Company. 1974. Aerial photographs on microfilm. The Kodak Compass 1974(2):[1-2].

By coordinating remote sensing reduction with reader magnification, engineering quality images are available on aperture cards.

56. Eastman Kodak Company. 1979. Preserving, restoring color transparencies. *Kodak Tech Bits* 1979(1):3-4.

The promise of a fifty year lifetime for color images. If the next fifty years are anything like the last fifty years, we need not plan beyond them anyway.

57 Edgerton, C.D. 1974. The Mine Map Repository--a source of mine map data. Washington: U.S. Bureau of Mines. Information Circular IC 8657. Reprinted in *Journal of Micrographics* 8(5):235-240. (NTIS PB-240 136/2ST)

Central source for mine map information, the MMR uses a computerized index for locations of mine maps, is currently filming 500,000 maps and is actively seeking permission to film all U.S. mine maps extant. Indexes of work accomplished are released periodically, both by the Bureau of Mines ICs 8379, 8453, 8519, 87779, and through NTIS.

58. Ehrenberg, Ralph. 1973. Reproducing maps in libraries and archives: the custodian's point-of-view. *Special Libraries* 64(1):18, 20-24.

Discussion of photographic reproduction, map accessibility and map handling by a strong advocate of map microfilming, and, at the time, Director of the Cartographic Archives Division, U.S. National Archives and Records Service. For companion article, see LaHood, 1973.

59. Ehrenberg, Ralph E. 1975. Cartographic archives: a perspective. Special Libraries Association, Geography and Map Division, *Bulletin* 99: 2-24.

An excellent introduction to the types of maps to be found in archives, how they are arranged, and an excellent bibliography of the literature. These holdings are, more and more, available on microfilm.

60. Ehrenberg, Ralph E. 1976. Photocopying rare maps and atlases. *AB Bookman's Yearbook* 1976:31-42.

Covers history and motives for photocopying (including microfilming) maps. Includes a thorough discussion of the National Archives 105mm program and an extensive set of bibliographical notes.

61. Ehrenberg, Ralph E. 1977. Conserving a cartographic heritage: microfilming at the National Archives of the United States. International Council on Archives, Microfilm Committee, *Bulletin* 6:21-27.

Discussion of the manuscript map microfilming project (105mm) at NARS, and the 70mm microfilm salvage of their historic aerial photograph collection. Subsequent to the article, NARS has added contrast enhancement to the latter program (personal communication).

62. English, J. 1960. Photo map volumes aid title searchers. *Industrial Photography* 9(4, Section 2): 44-45.

63. Engineering Graphics. 1977. 35mm or 105mm microfilm for your drawings? *Engineering Graphics* 17(5):

Brief discussion of the atributes of the respective formats. As in most cases where engineers are involved, maps and drawings are synonymous.

64. Eyton, J. Ronald. 1978. Macrophotography of satellite images. Photogrammetric Engineering and Remote Sensing 44(8):1019-1021.

LANDSAT 35mm and 70mm imagery enlargement from 1:3,369,000 to 1:60,000 using Kodak 2483 color film and photomicrography techniques.

65. Fair, Judy. 1979. Microforms management in special libraries: a reader. Weston, CT: Microform Review, Inc.

Assemblege of articles for general librarians. Includes maps (Chapter II), reprinting a review by Phil Hoehn of the WAML meeting on microcartography at UCSD, October 1977.

66. Fielstra, Gerritt E. 1950. Photoreproduction of maps: practical? Library Journal 75(6):464-465.

Conventional arguments on why microforms will not work for maps, beginning with lack of detail, inability to blow-back to original size, etc. These arguments parallel those of scientists in the 50's who thought that computers were only fast calculators, while Robert Oppenheimer argued they were fundamentally different things: the same is true of microcartography.

67. Flanders, D.W. 1977. The design and preparation of maps for 35mm projectors. The Cartographic Journal 13(1): 89-93.

Human factors considerations for group viewing. While a special case exercise, there are a number of observations pertinent to all projective map microforms.

68. Fortescue, Major F.S. 1968. The requirement of 70mm microfilm for map reproduction. In: National Reprographic Centre for Documentation, A technical appraisal of the 70mm format for map reproduction, pp. 1-9. Hatfield: Hertfordshire County Council, Technical Library Service.

Ordnance Survey experiments confirm the need for 70mm as the smallest practical format for large map microfilming.

69. Francis, D.G. 1975. Description of the conversion of the Department of Lands and Survey plan records to microfilm. Wellington: Department of Lands and Survey. Photoprocessed.

System design and cost/benefit analysis of an aperture card system. Includes discussion of the need for color, and the hope that it can be realised in the near future.

70. Functional Photography. 1979a. Aerial photography for civil engineering. Functional Photography 14(1):20-21.

35mm aerial photography can be cheaper, more timely and more relevant than any other type in certain applications.

1. Functional Photography. 1979b. Handling a hot potato! Functional Photography 14(2):20-25.

35mm infrared aerial photography used to detect crop diseases.

2. Gale, M.T., R.S. Engelbrecht, J. Kane, and K. Knop. 1978a. Surface-relief microimages. The Journal of Micrographics 11(3):155-162.

Though not strictly about maps, this article on zero order diffraction (ZOD) embossed microforms is too important to omit. ZOD's offer archival, high resolution color separations of maps and imagery, all by means of intentional diffraction of white light into selected spectromatic componants. The technology is ideal for topographic mapping, where large runs of a single color separation are needed; the separations can then be joined into complete, full-color maps at the map library level. Set up on a 105mm microform, such seperations would have full utility for large-size blow-back at every step of the map-use chain. What is wanting is a large-scale producer/user combination.

3. Gale, M.T., J. Kane, and K. Knop. 1978b. ZOD images: embossable surface-relief structures for color and black-and-white reproduction. Journal of Applied Photographic Engineering 4(2):41-47.

See Gale, et al., 1978a.

4. Gardiner, R.A. 1977. Sheet maps on microfilms? Geographical Magazine 50(2):143.

Ordnance Survey microfilm program.

5. Gear, James L., Robert H. MacClaren, and Mary McKiel. Film recovery of some deteriorated black and white negatives. The American Archivist 40(3): 363-368.

Salvage of 4x5 inch negatives using two different recovery procedures, both based on separating the emulsion from its original backing, flattening and then duplicating it.

6. Geoff, E.A., F.B. Shaffer, and R.M. Walchli. 1967. Summary report: Discussion Group 5, the assessment of map-display systems in aeronautics charts and map displays, In: J.J. McGrath (Ed.), A JANAIR Symposium. Goleta, CA: Human Factors Research Institute, Inc. (NTIS AD 661 490).

7. The Geographical Magazine. 1976. A geographer defends OS pricing policy. The Geographical Magazine 49(8):525.

Announces consideration by the Ordnance Survey of producing 1:50,000 topographic maps on 35mm slides and overhead transparencies.

8. Gibson, Stephen W. 1959. The use of microfilm in a proposed map viewing, reproduction, and distribution system. National Microfilm Association Proceedings 8:117-125.

Chief, Map Reproduction Section, Topographic Systems Branch, U.S. Army Engineer Research and Development Laboratories discussing a system whose components include 11,000 70mm map microforms per auto retrieval unit (stackable), a Target Map Coordinate Locator, and color printing from the map microform. Twenty years later, such a system is becoming increasingly cost justifible.

79. Goes de Azevedo, Renato. 1979. International cartographic rules for the demarcation of the earth's surface. Journal of Micrographics 12(5): 301-303.

Bibliographic control of earth related information, including maps, by means of microfiche coded in accordance with International Map of the World (IMW) 1:1,000,000 quadrangles, with standard breakdowns into larger scale subdivisions. Currently in use by Sensora Ltda. in Brazil for natural resource inventories COM microfiche.

80. Gold, Marvin H., Patrick F. Grosso, and Andrew A. Tarnowski. 1974a. Automated electron beam recorder cartographic/graphic arts typesetting system for production of color map products. Proceedings of the American Congress on Surveying and Mapping, Fall Convention, Washington, D.C., September 10-13, 1974:54-83.

Computer driven micromap system for the generation of 70mm and 4x5 inch images.

81. Gold, M.H., P.F. Grosso, and A.A. Tarnowski. 1974b. EBR extension of graphics generator to include symbols. [cover title: To add a graphic arts character generation capability to the present study of electron beam recorder (EBR) techniques for map production."] Stamford, CT: CBS Laboratories. (U.S. Army Engineer Topographic Laboratories ETL-CR-74-12; NTIS AD number unavailable.)

Continuation of Gold, et al., 1974a.

82. Goodwald, J.F. 1968. Storage and reproduction of mapping data. Journal of Micrographics 1(3):82-84. (NMA 68-0011)

Geodetic information on 35mm or 70mm aerial photographs mounted on 17 inch x 400 ft. rolls (28,000 35mm images) for 20 second retrieval.

83. Greed, James J. Jr., and Gerald N. Wallmark. 1977. Electron beam recorder applications study. Norwalk, CT: Carson Alexiow Corp. Report FF-12. U.S. Army Engineer Topographic Laboratories ETL-0120. (NTIS AD A048 666/2ST).

Mass memory COM cartographic products via EBR.

84. Greenburg, G.L. 1971. Irradiation effect on the perception of map symbology. International Yearbook of Cartography 11:120-126.

Concentrates on the effect of projected negative and positive images on comprehension, especially in regard to projected map displays (PMDs) on military aircraft. Suggests that maps read via projection may need modification.

35. Grosso, P.F., A.A. Tarnowski, and M.H. Gold. 1973. Investigation, experiments and study of Electron Beam Recorder (EBR) techniques for map production. Stamford, CT: CBS Laboratories. (NTIS AD-767 555/6)

R&D study of digitally generated micromap color seperations.

36. Grosso, P.F. 1974. Application of electron beam recording for automated cartography, In: International Conference on Automation in Cartography (Auto Carto I). Falls Church, VA: American Congress on Surveying and Mapping and U.S. Geological Survey.

Report on EBR color map separation generation on microforms.

37. Grosso, P.F., and A.A. Tarnowski. 1976a. Application of electron beam recording in graphics. Proceedings of the Technical Program, Electro-Optical Systems Design Conference, 1976, [and] International Laser Exposition, New York, New York, September 14, 15, 16, 1976: 347-352.

More on the microfilm/computer interface in mapping applications.

38. Grosso, P.F., and A.A. Tarnowski. 1976b. Cartographic electron beam recording. Paper delivered at the Second Annual William T. Pecora Memorial Symposium on Mapping and Remote Sensing Data.

39. Grosso, P.F., and A.A. Taranowski. 1976c. Electron beam recording on film--applications and performance. Electro-Optical Systems Design 8(8):56-61.

90. Grosso, P.F., and A.A. Taranowski. 1977a. Electron beam recorder (EBR) system. Fairfield, CT: Image Graphics, Inc. (U.S. Army Engineer Topographic Laboratories ETL-0111; NTIS AD-A044 401)

Generating maps on 35mm, 70mm, 105mm, and 5x8 inch film, the EBR addresses 32,000 x 32,000 points with a repeatability of 1/30,000 at a geometric fidelity of .03%. (From the abstract.)

91. Gunther, Alden C. 1974. A system for topographic inquiry. Number 1. Micrographic subsystem. Fort Belvoir, VA: U.S. Army Engineer Topographic Laboratories. (NTIS AD-923 480/8ST; NMA 74-0292)

Random access automatic retrieval microfilm map system.

92. Hanback, Hazel S. 1960. Photographing drawings and maps on 105mm film. National Microfilm Association Proceedings 9:104-106. (NMA 60-0015)

In spite of its title, this article does not treat maps specifically.

93. Harlow, Neal R. 1939. The facsimile card catalog. Library Journal 64:250.

Application of microprints to the card catalog.

94. Hasselblad. 1972? Aerial photography. Goteborg, Sweden: Victor Hasselblad Aktiebolag.

Series of articles illustrating use of various Hasselblad 70mm cameras for different application. Articles are entered separately in this bibliography.

95. Hirschsohn, Ian. 1977. Case studies of the dramatic rise of COM graphics. National Micrographics Association Proceedings 26:3-6.

Discussion of DISSPLA (Display Integrated Software System and Plotting Language) of COM maps by the Naval Weapons Center, China Lake, CA, and the Federal Energy Administration.

96. Hitchcock, Robert D. 1976. Seafloor mapping system. Official Gazette of the United States Patent and Trademark Office 949(15):2167. Patent 3,978,444.

Intriging invention intended to record stereo imagery of the seafloor on microfilm in a surface ship. Especially important is the use of fiber optic "funnels" and "staircases" which reduce the size of the images and could find useful application for projection equipment.

97. Hoehn, Phil. 1977. Maps in microform: small can be beautiful. CU News (General Library, University of California, Berkeley) 32(41):5-7.

Review of the Western Association of Map Libraries "microcartography" meeting, October 6-7, 1977, at the University of California, San Diego. Reprinted in Fair, 1979.(entry number 65)

98. Haldahl, Jeanne H., and Beard, Harold G. 1974. The National Geodetic Survey vertical publication data base. Proceedings of the American Congress on Surveying and Mapping, Fall Convention, Washington, D.C., September 10-13, 1974: 148-165.

Program to supply geodetic information on a microfilmed, quadrangle basis. Includes announcement of newsletter.

99. Honick, K.R. 1970. Developments in pictorial navigational displays. International Yearbook of Cartography 10:104-109.

Discussion of British developments of Projected Map Displays. Especially sensitive to the visual problems incumbent in the projection process. Also emphasizes the need for color microforms and redesign of map information based on human factors discoveries in PMD research. "...published in a longer version in The Cartographic Journal, December 1968."

100. Honick, K.R. 1971. Microphotography of maps on 35-mm colour film, In: Proceedings of the Third International Congress on Reprography, London, March 15-19, 1971: 156-159. Guildford, England: IPC Science and Technology Press Ltd

Based on work with navigational displays, discusses human factors, including the need for color films, cameras, optics, and the need for map design based on least common denominator factors discovered under adverse use situations, where clarity of information transmission is all important.

1. Ingersoll, Andrew P. 1974. Lunar studies. Pasadena, CA: California Institute of Technology. (NTIS/STAR N74-27349)

The creation of a rapid access microfiche library system of lunar photographs is discussed.

2. Information Dynamics Corporation. 1963. Study of NIMBUS satellite photographic archives system requirements for unitized microform. Wakefield, MA: Information Dynamics Corp. (NTIS PB-169 372)

System design study, including cost/benefit analysis, appropriate equipment and user needs.

3. Infosystems. 1977. A solution to the paper chase. Infosystems 24(10):114.

On site microfilming of real estate (plat) maps.

4. Irwin, M.M. 1976. Survey of digital image scanning systems. Fort Belvoir, VA: U.S. Army Engineer Topographic Laboratories. ETL-0087. (NTIS AD-A036 337/4GA)

"...devices which digitize data from film to use as input to a digital computer," i.e. Computer Input Microfilm (CIM) hardware survey (from the abstract).

)5. James, Dr. Wesley P. 1972? Hasselblad cameras aid in coastal pollution study, In: Hasselblad, Aerial Photography: 12-19. Goteborg, Sweden: Victor Hasselblad Aktiebolag.

Medium format (70mm) aerial photography.

)6. Jeffery, Margaret, Hazel O'Hare, and Christopher Board. 1975. Choropleth mapping on the microfilm plotter: an attempt to improve the graphic quality of automated maps. International Yearbook of Cartography 15: 39-46.

COM output on the Calcomp 1670 film plotter at the London School of Economics.

)7. Kansas. Legislative Council. Research Department. 1942. Microfilm for public records. Topeka: author. Publication No. 114.

Includes mention of San Diego, CA, assessment records.

)8. Karlsten, Evald. 1972? Swedish police records--with the Hasselblad 500EL, In: Hasselblad, Aerial Photography: 10-12. Goteborg, Sweden: Victor Hasselblad Aktierbolag.

Airborne road survellience with 70mm cameras.

109. Kassebeer, U. 1978. Flurkarten auf mikrofilm. Nachrichten aus dem öffentlichen Vermessungsdienst des Landes Nordrhein-Westfalen Heft 1, 1978: 8-17.

Cadastral map microfilming in Düsseldorf in a standard 35mm aperture card format.

110. Kelk, Brian. 1974. Hardware for automated cartography, In: British Cartographic Society, Automated Cartography: papers presented at the annual symposium of the British Cartographic Society, Southampton 1973. London: British Cartographic Society, p. 40, "No.3. Microfilm plotters."

111. Kidd, Betty. 1977. Maps in geneological research. Families (Ontario) 16(4):153-165.

Includes information on cadastral maps of Ontario Province available in the Public Archives of Canada on microfilm.

112. Klimek, Adam. 1965. Engineering test of target MAP coordinate locator. Fort Lee, VA: U.S. Army General Equipment Test Activity. (NTIS AD-471 250/1ST)

Early model coordinate access map microfilm viewing ensamble.

113. Knibiehly, Earle M. 1974. Minaturized map reproducables. Proceedings of the American Congress on Surveying and Mapping, Fall Convention, Washington, D.C., September 10-13, 1974: 98-102.

Review of U.S. Army Engineer Topographic Laboratories research on the Projection Plate Maker (PPM) 35mm and 70mm systems using the Latady Micromap Camera.

114. Knight, James W. 1976. Microforms and their application to map collection. Archivaria 1(1):98-100.

115. LaHart, Jack. 1976. State government special usage microforms. National Micrographics Association Proceedings 25:151-154. (NMA 76-0413)

35mm map microfilms in use by Illinois' Department of Highways and the Illinois State Arcives' land records and maps of 1450 abandoned coal mines.

116. LaHood, Charles G., Jr. 1973. Reproducing maps in libraries: the photographer's point of view. Special Libraries 64(1):19, 25-28.

Head of the Library of Congress's Photoduplication Service on the technical difficulties implicit in map microfilming. Concludes with the hope that future maps be designed with microfilming in mind.

117. Lamar, Jeannine V., and Roy H. Stratton. 1974. Digital weather data displayed in a color image format. Journal of Micrographics 7(6):271-278. (NMA 74-0024)

The Rand Corporation's Climate Dynamics Study COM mapping, progressing from symbols to grey tones to color portrayal of complex weather factors.

118. Langelier, Gilles. 1977. Maps on microfilm. The Archivist (Public Archives of of Canada) 4(6):4-6.

The 105mm map microfilming program of the National Map Collection PAC.

119. Laymon, R.S. 1966. Study of two image interpreter map displays: chip versus hard copy. Santa Monica, CA: Systems Development Corporation. Technical Research Note 169. (NTIS AD-639 126)

120. Lawler, Bob. 1972. PMDS (Projected Map Display System) has arrived. TAC Attack June 12, 1972:24-25.

Navigational map microform display system for tactical aircraft.

121. Lewis, Elizabeth Matthew. 1976. Visual indexing of graphic material. Special Libraries 67(11):518-527.

Microfiche versus Super 8 film versus television cassettes as visual indexes for a slide collection. Thorough analytical presentation of cost factors and formulas, human factors and potential. Suggests such indexes for map collections (p. 526).

122. Lukes, George E. 1978. Mapping applications of optical power spectrum analysis. Eighth Annual Automatic Imagery Pattern Recognition Symposium, Proceedings on Emerging Patterns in AIPR:202-204.

Image analysis of 70mm cartographic microforms using computers. Bibliography.

123. Luxenberg, H.R. 1966. Cartographic source material storage, retrieval, and output system requirements for ACIC/IDHS. Los Angeles, CA: Planning Research Corporation. (Rome Air Development Command Technical Report RADC-TR-66-50; NTIS AD-484 259/7ST)

70mm fully automated subsystem for a map library.

124. Luxenberg, H.R., and R.L. Kuehn, Eds. 1968. Display system engineering. New York: McGraw-Hill.

125. Maas, Keith A. 1975. Earth resources "browse" microfilm--a challange for the microfilm industry. National Micrographics Association Proceedings 24: 322-327.

EROS Data Center LANDSAT imagery, Johnson Space Center color imagery, Kodak 7271 color film (250 L/mm resolution) and 7381 (500 L/mm) are discussed in relation to visual indexes.

126. McAdoo, Patrick G. 1973. A-7D PMDS (Projected Map Display Set). USAF Fighter Weapons Review, Winter 1973:15-17.

127. McCormack, Duncan. 1977. The cartograper of the future. New Zealand Cartographic Journal 7(2):6-10.

Brief mention of the microfilming of cadastral maps in New Zealand.

128. McDermott, Jack C. 1975. Local, county and state government--Missouri State Land Survey Repository. National Microfilm Association Proceedings 24:421-428.

129. McDonnell, Michael, M. 1976. Holographic terrain displays. Ft. Belvoir, VA: U.S. Army Engineer Topographic Laboratories. ETL-0083. (NTIS AD-A035 154/4GI)

Holographic stereomodel research and development.

130. McGregor, Dr. H.H., Jr. 1975. Vesicular photography--an overview. The Journal of Micrographics 9(1):13-20. (NMA 75-0102)

Discusses the qualities of the vesicular process, mentioning its potential for remote sensing microforms because of its grainless, high resolution nature (p. 19).

131. Magnan, G.A. 1969. Streamlining San Diego's County maps. Engineering Graphics 9(9):11-13.

132. Magnan, George A. 1975. Coordinating a region's cartography pays off for San Diego. Engineering Graphics 15(2):11-16.

Use of 105mm negatives as integral to the updating of maps by "scissors" drafting.

133. Maksimov, N.P., and F.V. Sidorov. 1970. Mikrofil'mirovanie kart i certezej. Moscow: NEDRA. Translated as: Microfilming of maps and drawings (selected parts)[ch. 1, 2, 4, 7, 9 and bibliography]. Eric Peabody, translator. U.S. Army Foreign Science and Technology Center. Report No. FSTC-HT-23-1461-73. (NTIS AD-770 522/1)

Parallel by the Russians of Western European and American microfilming. Extensive bibliography of Eastern European and Russian sources.

134. Maling, D.H. 1966. Some thoughts about the miniturisation of map library contents. The Cartographic Journal 3(1):14-17.

35mm microfilm for libraries in Great Britain. Includes discussion,(15-17) by R.A. Skelton, E.J. Parsons, H.A.G. Lewis and Brigadier R.A. Gardiner of this and a previous paper on map collections.

135. Markham, Robert P. 1977. Topographic maps on microfiche. Journal of Micrographics 11(5):315-318.

University of Northern Colorado's stripped-up 35mm program for 7.5-minute and 1:250,000 scale USGS maps.

136. Maugh, Thomas H. 1978. Holographic filing: an industry on the verge of birth. Science 201(4):431-432.

COM and CIM map research and development at the U.S. Army Engineer Topographic Laboratories. 105mm digital storage is 1.5 million bits versus a hologram's 200 million bits storage capacity. Predicts a two year gestation for simple storage and display but leaves unknown the time before such storage is used to generate printing plates. Does not consider the use of such data stores in-house by map collections individually. Illustrated.

137. Meadows, John M. 1976. Micrographics; double edged sword--source document microfilming and Computer Output Microfilm. National Micrographics Association Proceedings 25:118-123. (NMA 76-0404)

COM maps.

138. Menmuir, P. 1977. The automatic production of weather maps, In: British Cartographic Society, Automated Cartography: Papers Presented at the Annual Symposium of the British Cartographic Sociaty, Southampton 1973: 86-99. London: British Cartographic Society.

COM isopleth maps at the British Meteorological Office.

139. Meyer, Morton A., Frederick R. Broome, and Richard A. Schweitzer, Jr. 1975. Color statistical mapping by the U.S. Bureau of the Census. The American Cartographer 2(2):100-117.

Color COM maps from a statistical base.

140. Mier, F. 1969. 105mm in Europe. Engineering Graphics 9(9):20-22.

Spanish commercial microfilming service which includes map microfilming work in its repertoire.

141. Mikhail, Edward M., and Maurice K. Kurth. 1971. Holographie für topographische kartierungen. Bildmessg. und Luftbildwesen 39(4):176-177.

142. Milhollen, Hurst. 1946. Pictures invade the catalog: two examples of illustrated cards. Library Journal 71(11):803-804.

35mm contact prints on catalog cards.

143. Miller, Max L. 1974. Perpetuation and consolidation of Federal Survey Field Notes. Proceedings of the American Congress on Surveying and Mapping, Fall Convention, Washington, D.C., September 10-13, 1974:287-290.

U.S. Bureau of Land Management 35mm based system.

144. Mitchell, Peter A. 1976 Aerial ice reconnaissance and satellite ice information microfilm file. Washington: U.S. Naval Oceanographic Office. (NTIS AD-A037 300/1ST)

Historical ice data files and the program to locate relevant microfilms.

145. Mitchell, Peter A. 1977. Aerial ice reconnaissance and satellite ice information file--1976. Supplement 1. Washington: U.S. Naval Oceanographic Office. (NTIS AD-A043 046/2ST)

Reports all microfilm imagery of ice data added to the file in 1976. See: Mitchell, 1976.

146. Morrill, Philip. 1967. Prototype microfilm projector for navigational charts. Fort Belvoir, VA: U.S. Army Engineer Research and Development Laboratories. AERDL-1906. (NTIS AD-660 337)

The Projected Map Display (PMD) evaluation, and prediction of future prospects.

147. Mueller, Peter F. 1969. Standard microfilms for recording color multiple images. National Microfilm Association Proceedings 18:76-89. (NMA 69-0030)

The need for color, stereo microforms is obvious. Unfortunately, multiplexing's light loss precludes its use at present.

148. Muller, Jean-Claude, and John L. Honsaker. 1978. Choropleth map production by facsimile. The Cartographic Journal 15(1):14-19.

COM manipulation on 35mm film using a Tektronix plotter.

149. Murphy, Lawrence P., and Edward G. Trelinskie. 1976. Near real-time generation of 3-D terrain displays and profiles. Army Research and Development 17(4): [33].

Truly amazing combination of microfilm and computer: troops in the field TV monitor the local scene, it is identified in the headquarters computer terrain profile data base, regenerated from the data base in the form of a topographic model, manipulated to any orientation desired and plotted on microfilm as a synthetic scene. Every map collection should have one! Includes illustration.

150. Myhre, Richard J. 1973. Producing high quality negatives from ERTS black and white transparencies. Berkeley, CA: U.S.D.A. Forest Service, Pacific Southwest Forest and Range Experiment Station. USDA Forest Service Research Note PSW-287.

Photographic enhancement of 70mm images.

151. National Microfilm Association. 1970. Micropublishing in color: an interview with Harvey Hament and Norman Ginsberg. The Journal of Mocrographics 5(2): 93-96.

Costs for color are three times that of black-and-white microforms. Mentions filming navigational charts in color.

152. National Micrographics Association. 1979. Micrographics Index. Silver Spring, MD: author.

Annually updated, this is the best source of information on microforms there is. Actively solicits listings, multi-indexes them and makes copy available in micro or paper format at nominal cost. Those interested in microcartography should exploit the service this index provides.

153. National Reprographic Centre for Documentation. 1968. A technical appraisal of the 70mm format for map reproduction. Hatfield, Hertfordshire, England: Hertfordshire County Council, Technical Library Service.

Landmark series of five papers discussing the qualitative need for 70mm as a minimum format size in cartographic work. Includes discussions of all pertinent equipment factors based on British experience.

154. National Reprographic Centre for Documentation. 1972. Recommendations of methods of microfilming: maps and plans. Hatfield, Hertfordshire, England: author.

Aperture card systems design for map microfilming at 35mm.

155. Oman, Ray, and Charles Taylor. 1972. Map microfilming feasibility study. Washington: U.S. National Archives and Records Service, Cartographic Archives Division. Photoprocessed.

A very complete evaluation of all factors leading to the NARS decision to microfilm manuscript maps at 105mm. An important study which deserves to be published.

156. Olson, Judy M. 1976. Noncontiguous area cartograms. The Professional Geographer 27(14):371-380.

Another potential application for map microforms, taking advantage of size change by optical means, it would require a zoom lens on a reader/printer so that proportionate size could be controlled, or the use of a Zoom Transfer Scope with a Polaroid camera attached.

157. Optronics International Incorporated. 197-a. Digital color film scanning and writing/plotting systems, C-4000 series. Chelmford, MA: author.

COM and CIM in color. One of the best arguments for the long life utility of 105mm map microforms, both as input and as output.

158. Optronics International Incorporated. 197-b. High-speed rotating drum digital scanning and writing systems. Chelmsford, MA: author. Brochure No. 23.

COM, CIM, remote sensing capable equipment.

159. Optronics International Incorporated. 1978. Optronics introduces high-speed, simultaneous red-green-blue color seperation scanner system. The Optronics Journal 6:5.

Good argument for reducing map color separations to 105mm.

160. Ostby, Frederick P., Jr. 1968. Ballistic winds study. Hartford, CT: Travelers Research, Inc. (NTIS AD-668 922)

Project generating COM wind maps.

161. Osterhoff, W.E., W.K. Earl, and J.J. McGrath. 1966. Geographic orientation in aircraft pilots: achromatic display of color coded charts. Goleta, CA: Human Factors Research, Inc. Technical Report 751-8. (NTIS AD-648 331.

Human factors research on projected map displays for aircraft navigation.

162. Parker, D.J. 1960. New tools for microreduction and reproduction. National Microfilm Association Proceedings 9:158-164.

Discusses those high technologies only now being implemented: fiber optics (30-50x optical improvement), high resolution TV (20,000 lines per raster), RCA fine color electrofax map printer developed for the U.S. Army 70mm map color seperations and fiber optics image convertors.

163. Parker, D.J., and F.C. Myers. 1960. An electrostatic color map printer. Journal, Society of Motion Picture Television Engineers 69:744-748.

Regeneration of color maps from 70mm color seperations.

164. Parker, J. Carlyle. 1978. Michigan county and regional histories and atlases. Microform Review 7(6):349-350.

Review of a microset containing 81 atlases, 1 gazatteer and 9 plat books, from the period 1894-1931, at a cost of $6.00 per title, all on 35mm roll film. A model for all states/countries, with one reservation: they would be more useful on microfiche issued as individual titles.

165. Poole, Frank. 1965. Microfilming large drawings. American Archivist 28:584-585.

35mm and 105mm map microfilming in the city of Baltimore.

166. Potash, Lawrence M., and Thomas E. Jeffrey. 1978. Factors in design of hardcopy topographic maps. Alexandria, VA: U.S. Army Research Institute for the Behavorial and Social Sciences. Technical Paper 284. (NTIS AD-A049 621)

Includes study based on use of map microforms (p. 20), but more importantly it give empirical substantiation for the design of maps for microform mode. Very important bibliography on human factors design.

167. Rase, W.D. 1974. Kartographische darstellung dynamischer vorgänge in computer generierten filment. Kartographische Nachrichten 24(6):210-215.

COM maps.

168. Redstone, J. Lilian, and Francis W. Steer. 1953. Local records: their nature and care. London: G. Bell.

Recommends the substitution of photostats for maps and filming for preservation.

169. Rhind, David. 1974. High speed maps by laser beam. The Geographical Magazine. 46(8):393-394.

Discussion of the principles of direct plotting of computerized maps on various film bases, including full-size mylars and various reductions. Concentrates on the capabilities of laser plotters in terms of general types and sophistication in regards to resolution and continuous tone capabilities.

170. Richardson, Steven L. 1978. Remote sensing on a shoestring. Photogrammetric Engineering and Remote Sensing 44(8):1027-1032.

Shows how to construct a practical additive color viewer for less than $20. While based on the use of prints, the same techniques will work for transparent, micromaterials.

171. Rosenbluth, Roz. 1977. Pelham mapmakers. New York Times, Sunday, October 2, 1977, XXII, 4:5.

Use of microfilm in the current mapping program of the Sanborn Map Company. According to a later, personal communication, Sanborn is once again expanding its city mapping, this time on microfilm.

172. Rosenfeld, Azriel. 1979. Survey, picture processing: 1978. Computer Graphics and Image Processing 9:354-393.

Bibliography on computer manipulation of graphic media, much of it relevant to cartography. Section "D.1" is "Input, storage, nondigital processing;" section "C.4" is "enhancement and restoration," the former for microfilm relevant entries and the latter for the improvement of same.

173. Salchow, W. 1936. Karteikarte mit losen kleinbildern oder filmen. IID Communications 3[1]: col. sa. 1-2. [IID is Institute International de Documentation, later, Federation Internationale de Documentation]

Discusses reformatting roll film to unitized carriers. What is remarkable is that the basic formats for aperture cards and microfiche have been around so long.

174. Sarno, Frank J. 1972. The potential business uses of computer animation. National Microfilm Association Proceedings 21(II):178-184.

Mentions dynamic mapping (kinetic cartography?) (II:181-182).

175. Scheps, Bernard B. 1977. Automated processing of geographic information in image data forms. Fort Belvoir, VA: U.S. Army Engineer Topographic Laboratories. ETL-0114. (NTIS AD-B024 504/2L).

Reviews the present state-of-the-art and suggests lines of development for image data and slope map materials. Restricted distribution.

176. Schiffer, W. 1967. Der mikrofilm im vermessungswesen (Microfilming in Geodesy). Vermessungstechnik 15(8):302-304. Cited in LaHood, 1973.

177. Schmid, Calvin F., and Stanton E. Schmid. 1979. Handbook of graphic presentation. 2nd ed. New York: John Wiley & Sons. A Ronald Press Publication.

Chapter 11 is "Role of the computer in graphic presentation," with a section (pp. 291-297) on micrographics which discusses the U.S. Bureau of the Census program of COM maps, followed by a section on "Genigraphics," a General Electric propriatary computer output slide program capable of producing maps.

178. Schröder, C., and K. Stephan. 1974. Einheitliches mikrofilmsystem für kartographische zwecke. Militartechnik 1974(4):168 ff., and Vermessungstechnik 24(1):13-29.

Uniform microfilm system for cartographic applications.

179. Schweitzer, Richard H. 1973. Mapping urban America with automated cartography. Washington, D.C.: Government Printing Office. U.S. Bureau of the Census. Cited in Schmid, 1979.

180. Schweitzer, Richard H., Jr. 1974. Computer mapping with micrographics, In: Geographic base (DIME) file system: a forward look; conference proceedings, April 16 and 17, 1974, Boston, Mass:113-121. Washington: Government Printing Office. U.S. Bureau of the Census Computerized Geographic Coding Series GE 60 No. 5.

Explanation of the Census Bureau's 35mm-105mm COM map output, with emphesis on the computer mechanics. Includes discussion of further possibilities.

181. Schweitzer, Richard H., Jr. 1975. Micrographics: a new approach to automated cartography and graphics, In: Proceedings of 22nd International Technical Communication Conference, Anaheim, California, 1975: 305-316. Cited in Schmid, 1979.

182. Sibert, John L. 1978. Mapping offshore oil leases. Proceedings of the American Congress on Surveying and Mapping, 38th Annual Meeting, Washington, D.C., February 26th-March 4th, 1978: 128-134.

Includes "map production" (p. 132) as part of a COM black-and-white and color recorder system which yields 35mm slides.

183. Simmonds, M.C. 1976. Some aspects of the use of photography in map production in the Ordnance Survey. The Cartographic Journal 13(1):68-71.

#7 is "Microfilm Techniques" (p. 71). Includes 35mm and 70mm film materials used in map editing, updating and production.

184. Simmons, John. 1978. Micrographics in the National Cartographic Information Center. Special Libraries Association, Geography and Map Division Bulletin 112:47-50.

Numerous innovative visual indexing and historic map programs, as well as the more standard computer output index listings. The visual programs are limited to 35mm techniques.

185. Simmons, John. 1978? A description of the topographic quadrangle map series on microfilm. Reston, VA: U.S. Geological Survey, National Cartographic Information Center. Typescript.

35mm black-and-white roll program to film all of the 7.5- and 15-minute quadrangles of the USGS.

186. Simpson, J. Wesley. 1969. The ultimate in data bank storage and retrieval for aerospace imagery and documentation. National Microfilm Association Proceedings 18:90-98.

Calls for the establishment of a real-time graphic file of all space generated imagery, starting with a capacity of 25,000,000 scenes, 5 second access, multiple visual/print output, remote service, and ability to handle all formats from 35mm to 9 1/2 inches (241mm).

187. Slator, P.N. 1977. Comparison of photographic and digital imagery from film and solid-state-array remote sensing cameras. Photographic Science and Engineering 21(4):198-203.

"The film camera is preferred for cartographic...applications because the visual resolution limit is...higher...." (from the abstract). While the paper is not discussing microfilms, the qualitative considerations are relevant. At the same time, it should be noted that solid state arrays have no moving parts, and can be expected to improve in quality.

188. Smith, D. COM generated mapping. Paper delivered at COMTEC (Computer Micrographic Technology), 2nd Semi-annual General Meeting, October 6-8, 1971, San Francisco, California. Cited in Journal of Micrographics 5(3):155.

189. Society of American Military Engineers. 1968. Aeronautical Chart and Information Center: rapid projection ticking. The Military Engineer No. 397:374.

Use of 70mm in the production of aeronautical charts.

190. Society of American Military Engineers. 1979a. DMAHTC supports Army aviation. The Military Engineer No. 459:42-43.

Microfilm Projected Map Display System (PMDS) produced by the Defense Mapping Agency Hydrographic/Topographic Center.

191. Society of American Military Engineers. 1979b. New horizontal control publication system. The Military Engineer No. 461:269.

National Ocean Survey's Office of the National Geodetic Survey automated program of horizontal control information on microfilm.

192. Society of American Military Engineers. 1979c. LANDSAT imagery used to update hydrographic charts. The Military Engineer No. 463:[np].

Use of 70mm imagery in a combined color additive viewer/Zoom Transfer Scope apparatus to update nautical charts.

193. Spencer, Raymond D. 1978. Map intensification from small format photography. Photogrammetric Engineering and Remote Sensing 44(6):697-707.

35mm and 70mm. Bibliography.

194. States, Hugh W. 1961. Microfilm activity on the West Coast. National Microfilm Association Proceedings 10:42

Includes discussion of the California Department of Water Resources 35mm microfilms of land use mapping and its supporting documentation.

195. Statham, Ian. 1976. Photographic filtering for extracting orientation and aspect data for maps. Area 8(3):209-212.

Though not about microforms per se, the manipulative ideas apply.

196. Stauffer, Alvin P. 1938. Microphotography in the National Park Service. The Journal of Documentary Reproduction 1(2):172-177.

Discusses program for microfilming records pertinent to Historic Park Sites, including maps (p. 175).

197. Steiner, Dieter, and Anthony E. Salerno, Coauthors-editors. 1975. Chapter 12. Remote sensor data systems, processing, and management, In: Janza, Frank J., Editor-in-Chief, Manual of Remote Sensing. Vol. 1. Theory, Instruments and Techniques. Falls Church, VA: American Society of Photogrammetry.

Includes subsection on the use of microforms in remote sensing work, including their information carrying capacity both as input and output.

198. Steiner, Dieter, and Michael E. Kirby. 1977. Geometrical referencing of LANDSAT images by affine transformation and overlaying of map data. Photogrammetria 33(2):

Placing geographic coordinate controls on 70mm imagery.

199. Stephan, K, and C. Schröder. 1976. Das Einheitliche mikrofilmsystem für kartographische zwecke. (Centralized microfilm system for cartographic purposes.) Vermessungstechnik 24(1):13-14, 27-29.

Description of current equipment available for the microfilming and microduplication of cartographic records, i.e. German equipment with sufficient resolution and copyboard capacity.

200. Stevens, G.W.W. 1968. Control of line width in microfilm, In: National Reprographic Centre for Documentation, A technical appraisal of the 70mm format for map reproduction: 21-30. Hatfield, Hertfordshire, England: Hertfordshire County Council, Technical Library Service.

1. Stevens, Stanley D. 1970. Color microfilming of Sanborn maps for a local history collection. Western Association of Map Libraries Information Bulletin 1(1):2-12.

Thorough description of a landmark do-it-yourself project, including assembly of original documents, results of films used and costs associated with the finished product.

2. Stevens, Stanley, D. 1971. Microfilming maps in color. National Microfilm Association Proceedings 20: II48-II50. (NMA 71-0049).

See: Stevens, Stanley D. 1970.

3. Stockwell, Guy N. 1977. Reproducing topographic maps and navigational chart products utilizing the electrostatic image reproducer. Proceedings of the Annual Meeting of the American Congress on Surveying and Mapping, Washington, D.C., February 1977: 35-43.

Report on the U.S. Army's experiments with the regeneration of full-size maps from 70mm color seperation microforms.

4. Summers, Lionel. 1941. Microphotography keeps easements safe. Electrification News 6:13.

5. Taschioglou, Kemon P. 1957. The Polaroid transparency system. Special Libraries 48:426-427.

Instant, unitized microforms of 2 1/4 x 2 1/4 inches (70mm) and 3 1/4 x 4 inches (85x105mm). Film is available in black-and-white only. Probably an ideal approach for work in the library.

6. Taubes, Earnest P. 1973. Map microfilming project: methods and equipment for a proposed 105mm system. Washington, D.C.: U.S. National Archives and Records Service, Cartographic Archives Division. Photoprocessed.

The very thoroughly researched study on which NARS based its decision to use 105mm for manuscript map microfilming.

7. Taubes, Earnest P. 1979. Color microfilming of large documents. Journal of Micrographics 12(5):285-288.

Currently available 35mm and 105mm color microfilms suitable for map reproduction. Well researched and written. Essential knowledge.

08. Taylor, R.M. 1973. Human factors in maps and charts, a bibliography. London?: RAF Institute of Aviation Medicine. Report No. 543.

597 references assembled with Projected Map Display microform research in mind.

09. Taylor, R.M. 1974a. An evaluation of an experimental map for the Projected Map Display. London?: RAF Institute of Aviation Medicine. Report No. 545.

210. Taylor, R.M. 1974b. Subjective assessment of 1:250,000 scale maps for low altitude high speed navigation. London?: RAF Institute of Aviation Medicine. Report No. 544.

Projected map display research.

211. Taylor, R.M. 1975. Information theory and map evaluation. International Yearbook of Cartography 15: 165-181.

Summation of Projected Map Display human factors research as it applies to information transfer of topographic information. Includes results of tests designed to measure effectiveness of standard, special and projected maps. Extensive bibliography. Projected Map Displays used were 35mm.

212. Thornton, Frank R. 1978. Microfilm in Baltimore County. The Journal of Micrographics 11(4):283-285.

Complete, working 35mm map microform system, tied to a computerized indexing retrieval grid. Also discusses reader/printers. Illustrated.

213. Uhrin, D.C. 1977. Correlating microfilm mine maps with topographic maps. Washington, D.C.: U.S. Bureau of Mines and Energy Research and Development Administration. Bureau of Mines Information Circular IC 8762. (NTIS BM-IC-8762)

214. U.S. Army Engineer Topographic Laboratories. 1972-. Bibliography of in-house and contract reports (and annual supplements). Fort Belvoir, VA: USAETL. (NTIS AD-877 653L; AD-890 066L; AD-905 548L; AD-B005 275L; AD-B010 642L; AD-B019 966L; supplement 6, 1978, did not have an accession number yet. Note that documents with a "B" at the beginning were "restricted" at the time of issuance.)

These bibliographies constitute an unbroken chain in the development of map microreduction technology, including integration with computers, use of Electron Beam Recorders, lasers and holographic techniques for printing and/or regeneration.

215. U.S. Army Engineer Topographic Laboratories. 19--. ETL R&D activities (fiscal year) .

Annual report of the Topographic Development Laboratory.

216. U.S. Bureau of Land Management. 1970. Land and survey records: record preservation and service through microfilm. Denver?: BLM.

Preservation program for all land related records of the Federal surveys of the public lands, mainly those in the West. 35mm.

217. U.S. Department of Defense. 1962. MIL-P-9879A: photographing of construction/architectural drawings, maps and related documents, 105mm, requirements for. Philadelphia, PA: Naval Publications and Forms Center. (Commanding Officer, Code 105, Naval Publications and Forms Center, 5801 Tabor Ave., Philadelphia, PA 19120.)

218. U.S. National Aeronautics and Space Administration., Technology Utilization Office. 1972. Photography equipment and techniques: a survey of NASA developments. Washington, D.C.: Government Printing Office. NASA Special Publication SP-5099.

Concentrates on the 70mm format in a remote sensing context.

219. U.S. National Aeronautics and Space Administration. 1976. New method for developing clearer photos. NASA Activities 7(9):17.

Autoradiography (radioactive chemical) contrast enhancement of films/photos can be used in normal photo labs.

220. U.S. National Aeronautics and Space Administration. 1977a. More on how image enhancement may explain past events. NASA Activities 8(11):17-18.

Photo-based computer enhancement, in this case the subject is the "Shroud of Turin." Enhancement technique appears equally valid for faded maps.

221. U.S. National Aeronautics and Space Administration. 1977b. Restoring old photos --one of many uses of autoradiography processes. NASA Activities 8(12):24.

Image enhancement using radioactive chemicals on underexposed or faded photographic transparencies. "The two drawbacks right now are the high cost of the technique and the lack of an automatic processing method."

222 U.S. National Aeronautics and Space Administration. 1978. Making the invisible visible. Spinoff (an annual report) 1978:92-95.

The space-developed combination of filtering and 'contrast stretching' produces the sharpest contrast possible with today's equipment."

223. U.S. National Archives and Records Service. 1970. The selection and preparation of records for publication on microfilm. Washington, D.C.: NARS. Staff Information Paper 19.

224. Veaner, Allen B. 1964. European report: Reprography Congress, document reproduction activities in French and British libraries. Library Resources and Technical Services 8(2):199-204.

British Library proposal to make map aperture cards (p. 202).

225. Wertz, J.A. 1961. A system for microfilming maps. National Micro News 54:117-121.

226. White, Arthur Walmesley. 1970. "Printing on demand" and other Ordnance Survey developments. International Yearbook of Cartography 10:64-72.

The OS Survey Information on Microfilm (SIM) and large scale maps (1:1250 and 1:2500) on aperture cards. Of particular interest is the use of retail agents to supply paper prints from the microforms

227. Williams, Forest L. 1976. Micrographic projects at NARS. National Micrographics Association Proceedings 25:173-179. (NMA 76-0419)

Includes, in addition to their regular map microfilming program, mention of the Federal Government Micrographics Council, its monthly news bulletin and the role of affiliate members.

228. Wilson, John P. 1979. Map and chart information storage and retrieval. Proceedings of the American Congress on Surveying and Mapping, 39th Annual Meeting Washington, D.C., March 18-March 24, 1979:390-396.

National Cartographic Information Center (NCIC), map and chart information system, to which NCIC is keying its map microforms. Coordinate access.

229. Wilson, Leonard S. 1949. Lessons from the experience of the Map Information Section, OSS. Geographical Review 39(2):298-310.

Includes "Microfilming of Map Catalogues and Maps," (pp. 300-302) which details the OSS "project to microfilm all usable maps housed in the major collections of the United States." It should be noted that these microfilms were never deposited in the National Archives after the war and their current whereabouts is unknown. See also, entry number 46.

230. Wolter, John A. 1963. Rare maps and visual aids. Submitted to Professor Neville Pearson as a 3 credit starred paper in partial fulfillment of the requirement for EDCI 205, and the Degree of Master of Arts in Library Science. Typescript.

Thorough discussion of the methods and applications of 35mm slides and microfilm in a map library environment, as well as 3 x 5 inch graphic catalo cards, reader/printers and associated technologies.

231. Woods, Erik K. 1971. Micromap camera for display systems. Fort Belvoir, VA: U.S. Army Engineer Topographic Laboratories. (NTIS AD-738 163)

Camera for source material to 75 x 75 inches yielding microfilm from 35 x 60mm to 9 x 9 inches.

232. Zech, R.G., L.M. Ralston, B.R. Reddersen, and H.N. Roberts. 1977. Microreduction of graphic information study (MEGIS). Melbourne, FL: Harris Corporation, Electronic Systems Division. (U.S. Army Engineer Topographic Laboratories ETL-0063; NTIS AD-A051 532).

State-of-the-art 70mm based 20 x 20mm holographic reductions of map separations. Includes complete system needs and technical results, possible microfiche (105mm) carriers, automatic retrieval and projection considerations. Very important reference citations after each chapter. A landmark work.

233. Zech, R.G. [n.d.]. Digital maps on microfiche. Rome, NY: Rome Air Development Command?

Microfilm mass memory systems, using holographic recording on COM fiche: 30 million bits per microfiche.

Index to the Bibliography

NOTE: Numbers refer to entries in the Bibliography.

Aberdeen City Library, 44
Aerial photography, see remote sensing
Aeronautical charts, 6, 50, 90, 120, 126, 189, 208, 209, 210
Animation, 42, 174
Atlases, 2, 20, 24, 106, 139, 179, 180
Autoradiography, 219, 220
Archives: Canada, Langelier, this volume, 111, 118; France, 48; Hungary, 19; Harrison, Lundquist, this volume; New York, 28; Spain, 43; U.S., Taylor, this volume; 58, 59, 61, 155, 216, 223, 227
Automatic retrieval, 78, 82, 91, 101, 112, 123, 186
Baltimore:(city), 165; (county), 212
Bibliographies: 4, 60, 133, 144, 145, 152, 197, 208, 214; map design, 166; picture processing, 193
Bibliographic control: Fraim, Georgenson, Greenberg, Kimball, Markham, Wolter, this volume; 18, 29, 34, 44, 78, 79, 91, 93, 121, 142, 228.
Brazil, 79
Browse files, 125
British Library, 224
British Meterological Office, 138
California Department of Water Resources, (land use) 194
Cameras: Lataday, Langelier, Lundquist, Taylor, this volume; 113, 218, 231
Canada: Langelier, this volume; 5, 111, 118
Chicago, 35
Color: Fraim, Lataday, Lundquist, Markham, Talbot, Taylor, this volume; 1, 9, 13, 15, 20, 24, 28, 31, 38, 51, 52, 56, 60, 64, 69, 71, 72, 73, 78, 80, 85, 86, 87, 88, 89, 90, 117, 125, 135, 139, 147, 151, 155, 157, 158, 161, 170, 192, 197, 201, 206, 207, 232
Computers, 3, 8, 9, 16, 20, 21, 24, 42, 80, 81, 95, 104, 106, 149, 175, 177, 179, 186, 222, 233
Computer Input Microfilm (CIM), 3, 37, 104, 122, 136, 157, 158, 159, 197, 233 (digital store), see also Mass memory,
Computer Output Microfilm (COM), 8, 9, 12, 16, 20, 21, 24, 33, 37, 39, 42, 50, 80, 81, 83, 87, 88, 89, 90, 95, 106, 110, 117, 136, 137, 139, 148, 149, 157, 158, 160, 167, 181, 182, 188, 197
Coordinates: Georgenson, Greenberg, this volume; 78, 79, 112, 198, 212, 228
Counties, 107, 131, 132, 164, 212
Critical studies, 13, 63, 66, 100, 102, 119, 153, 155, 166, 168, 200, 206, 207, 208, 209, 210, 211; cameras: Lataday, this volume, 36; Electron Beam Recorders: 89
Czechoslovakia, 11
Displays: Fraim, Georgenson, Greenberg, Kimball, Lataday, Markham, Taylor, Wolter, this volume; 1, 6, 13, 32, 67, 76, 84, 99, 120, 126, 146; design, 124
Electron Beam Recorder (EBR), 9, 21, 33, 80, 81, 83, 85, 86, 87, 88, 89, 90
Embossing, 72, 73
Fiber optics, 96, 162
Film: Lundquist, Markham, Taylor, this volume; 3, 9, 14, 64, 68, 75, 125; vesicular, 130; see also, color.

Format: 9; conversion (roll to unitized) 173;
35mm: Lundquist, Markham, Talbot, Taylor, this volume; 15, 22, 44, 46, 51, 52, 55, 63, 70, 71, 74, 77, 82, 99, 100, 109, 113, 115, 134, 135, 142, 148, 153, 154, 164, 165, 180, 182, 183, 184, 193, 211, 212, 226, 229, 230
70mm: Barkee, Fraim, Lataday, this volume; 21, 27, 31, 33, 36, 44, 61, 68, 80, 82, 94, 108, 113, 122, 123, 150, 153, 162, 163, 180, 183, 193, 198, 218
4"x5": 75, 80
105mm: Langelier, Lundquist, Talbot, Taylor, this volume; 51, 52, 53, 60, 61, 63, 92, 111, 118, 132, 140, 155, 165, 180, 207, 217
310mm: 20, 24
France, 48
General, 11, 26, 45, 78, 152, 168
General Electric, 177
Genealogy, 111
Geodesy, 82, 98, 176, 191
Germany (Federal Republic), 109, 167, 176, 178, 199
History, 60
Holography, 9, 10, 129, 136, 141, 232
Human factors, 13, 67, 76, 84, 99, 119, 121, 146, 161, 166, 208, 209, 210, 211
Hungary, 19
Ice, 144, 145
Illinois, 115
Image enhancement, 14; see also, autoradiography
Indianapolis, 54
Interactive, 149
Land records, 47, 51, 52; see also, real estate
Land use, 194
Lasers, 9, 10, 12, 169
Libraries: Georgenson, Hoehn, Kimball, this volume; 29, 44, 45, 46, 65, 66, 78, 93, 97, 123, 134, 135, 201, 205, 224, 229, 230; Library of Congress, 116
Local government: Kimball, Lundquist, this volume; 28, 35, 40, 49, 51, 52, 54, 107, 115, 128, 131, 132, 212
London School of Economics, 106
Los Angeles: Kimball, Lundquist, this volume;
Manipulation: Talbot, Wolter, this volume; 9, 14, 18, 21, 53, 56, 64, 132, 150, 156, 170, 175, 189, 192, 195, 198, 200, 213, 218, 219, 221; bibliography, 172
Map design, 13, 16, 166, 208, 209, 210
Mass memories, 83, 233; see also, computer input microfilm
Meterology, 117, 138, 160
Michigan, 164
Mine maps, 57, 213
Missouri, 128
National Reprographic Centre for Documentation (NRCd), 153
Navigation: Fraim, this volume; 120, 151, 161, 190, 208, 209, 210
New York City, 28, 51, 52
Oceanography, 96, 105
Ordnance Survey of Great Britain, 68, 74, 77, 153, 183, 226
Polaroid, 205
Products: (commercial), Barkee, Markham, Talbot, this volume; 45, 164
Projection, see display
Random access, 91
Royal Air Force (UK), 208, 209, 210

Real estate: Kimball, this volume; 45, 49, 51, 52, 62, 103, 107, 109, 111, 128, 143, 204, 216; see also, land records
Remote sensing: Greenberg, this volume; 8, 17, 21, 22, 27, 30, 55, 64, 70, 71 82, 94, 96, 101, 102, 105, 108, 119, 122, 125, 130, 144, 145, 150, 162, 163 170, 186, 187, 192, 197
Reprographics: Kimball, Lundquist, Talbot, Taylor, this volume; 116, 162, 163 203, 226
Research and Development, 9, 20, 24, 31, 33, 72, 76, 80, 81, 83, 85, 86, 87, 88, 89, 90, 129, 208, 209, 210, 211, 215
Restoration (of film), 56, 75
Retrieval, see bibliographic control
Saint Louis University, 28
San Diego, 107, 131, 132
Sanborn Map Company, 171, 201
Separations (color or feature): Greenberg, Latady, this volume; 9, 31, 33, 72, 162, 163, 170
Solid state arrays, 187
Sources, 45
Spanish Microfilm Service, 43
Standards: Stevens, this volume; 7, 23, 217, 223
Statistical maps, 20, 24, 106
Stereoptic effects, 129, 147, 149
Storage, 56
Systems, 78, 82, 186, 225
United States: Air Force: Fraim, this volume; 6, 126, 189
Army Engineer Topographic Laboratories, 9, 31, 33, 78, 80, 81, 83, 90, 104, 113, 129, 136, 146, 149, 162, 175, 203, 214, 215, 231
Bureau of the Census, 20, 24, 139, 179, 180
Bureau of Land Management, 143, 216
Bureau of Mines, 57
Defense Mapping Agency: Latady, this volume; 190
Defense Mapping Agency, Aerospace Center: Fraim, this volume
Federal Energy Administration, 95
Federal Government Micrographics Council, 227
Forest and Range Experiment Station (Berkeley), 1, 3
National Aeronautics and Space Administration, 218, 219, 220
National Archives, 58, 59, 60, 61, 155, 206
National Cartographic Information Center: Greenberg, this volume; 184, 228
National Environmental Satellite Service, 102
National Geodetic Survey, 98, 191
National Park Service, 53, 196
Naval Oceanographic Office, 144, 145
Naval Weapons Center (China Lake), 95
Office of Strategic Services, 46, 229
Union of Soviet Socialist Republics, 133
Vatican, 28
Westchester County, NY, 49
Zero Order Diffraction microforms, 72, 73
Zoom lenses, 1

WESTERN ASSOCIATION OF MAP LIBRARIES

Occasional Paper Series

No. 1: *Catalog of Sanborn Atlases at California State University, Northridge,* by Gary W. Rees and Mary Hoeber. 1973. xxi, 122p. (LC # 73-5773 ; ISBN 0-939112-01-9) $ 4.00

No. 2: *Union List of Sanborn Fire Insurance Maps Held by Institutions in the United States and Canada, Vol. 1, Alabama to Missouri,* by R. Philip Hoehn. 1976. xvii, 178p. (LC # 76-6129 ; ISBN 0-939112-02-7) $ 5.00

No. 3: *Union List of Sanborn Fire Insurance Maps Held by Institutions in the United States and Canada, Vol. 2, Montana to Wyoming; Canada and Mexico,* by William S. Peterson-Hunt and Evelyn L. Woodruff; *with a Supplement and Corrigenda to Volume 1,* by R. Philip Hoehn. 1977. xv, 201p. (LC # 76-2129 Rev. ; ISBN 0-939112-03-5) $ 6.00

Occasional Papers No. 2 and No. 3 (when ordered together) ISBN 0-939112-04-3) $10.00

No. 4: *Index to early twentieth century city plans appearing in guidebooks: Baedeker, Muirhead-Blue Guides, Murray, I.J.G.R., etc., plus selected other works to provide worldwide coverage of over 2,000 plans to over 1,200 communities, found in 74 guidebooks,* by Harold M. Otness. 1978. xxx, 94p. (LC # 78-15094 ; ISBN 0-939112-05-1) $ 6.00

No. 5: *The Maps of Fiji: a selective and annotated cartobibliography,* by Mason S. Green. 1978. xx, 70p. (LC # 78-24066 ; ISBN 0-939112-06-X) $ 4.00

No. 6: *Microcartography: applications for archives and libraries,* edited by Larry Cruse, with the assistance of Sylvia B. Warren. 1981. x, 200p. (ISBN 0-939112-07-8) (LC # 81-19718)

No. 7: *Index to nineteenth century city plans appearing in guidebooks: Baedeker, Murray, Joanne, Black, Appleton, Meyer, plus selected other works to provide coverage of over 1,800 plans to nearly 600 communities, found in 164 guidebooks,* by Harold M. Otness. 1980. xxiv, 84p. (LC # 80-24483 ; ISBN 0-939112-08-6) $ 6.00

No. 8: *Printed Maps of Utah to 1900; an annotated cartobibliography,* by Riley Moore Moffat. 1981. xvi, 177p. (LC # 81-659 ; ISBN 0-939112-09-4) $10.00

Standing orders for this Series, or single items, may be ordered from: Western Association of Map Libraries, c/o Stanley D. Stevens, Treasurer, University Library, University of California, Santa Cruz, CA 95064.

Members of WAML receive 40% discount off price shown; postage & tax added where appropriate. All other orders, List Price shown, post & tax.